THE
GRAVITY
WELL

THE
GRAVITY
WELL

America's Next, Greatest Mission

Stephen Sandford
with Jay Heinrichs

Gavia Books
831 Spruce Avenue
Pacific Grove, CA 93950
www.GaviaBooks.com

Interior design by Tessa Avila

FIRST EDITION
10 9 8 7 6 5 4 3 2 1

Library of Congress Control Number 2016948082

ISBN 978-0-9962422-9-5

Printed in the United States of America

Contents

*To **James E. Webb, Hugh L. Dryden** and **Robert C. Seamens, Jr.,** three of America's greatest heroes.*

Each man was impressive in his own right—power broker and advocate, aeronautics and flight system scientist, and consummate system engineer and executive—commanding his own spheres of expertise. It was only by working together that they created America's formidable space flight infrastructure and met President Kennedy's audacious challenge to land men on the Moon and return safely. They set the bar high for all of us who follow.

It's space. It's filled with chance, circumstance, and bad luck. It doesn't cooperate. At some point, I promise, at some point every single thing is gonna go south on you, and you'll think: this is it. This is how I end.

And you can either accept that... or you can get to work.

That's all it is. You simply begin.

Solve one problem. Then the next one, then the next.

You solve enough problems... and you get to come home.

Mark Watney in the film *The Martian*

Preface

Some 50,000 years ago, a man stood up and left his home fire.

A man stood up and left his home fire.

Think about the astonishing evolution and accomplishments in that one sentence. At that moment our species had been standing up for at least three million years, an ability that entailed some risk. A standing ape is visible and vulnerable, an easy target for predators and enemies. But the bold posture frees the hands for tools, and it extends the vision to the horizon.

For many millennia, men—or, more likely, women—had been making fires and cooking meat. Mastering fire must have taken heroic courage amid many painful mistakes. Yet the feat represents another key trait of humans: the willingness to transform hostile forces into allies. And so the man had the DNA of a seeker and an inventor. And now he left his home fire.

Accompanying him were his hunting partners, highly skilled marksmen and warriors who knew how to plan a hunt and patiently execute the strategy. Most of them carried the latest technology: a spear with a stone so sharp they could have shaved with it. But our man carried only a stone knife with a leather handle.

The men walked for hours until they came to where they knew a herd of deer liked to gather. They signaled to each other, walked in a noiseless crouch, and moved within spear-throwing distance of the herd. But just as the spear men stood up, one of the deer sniffed the air and raised the alarm, and the whole herd took off bounding through the woods. The spear men just stood there. They knew they didn't have a chance.

Yet our man took off after the deer, armed only with his knife. Though he was the fastest runner among the humans, the deer ran six times faster. What he lacked in deer-speed he made up in persistence. He kept running, long after he lost sight of his prey, following the tracks, jogging steadily, for ten miles until one of the older deer began to stumble and fall behind the others. The man caught up and leaped onto the animal's back.

Thousands of years after our man got his deer, one of his descendants once more wandered far from the home fire. The deer had long since disappeared, hunted out of existence within their known world. To make things worse, the climate had been changing. The winters were getting colder, and the summers failed to warm. In search of game, this man walked for days until he came to the shores of an icy sea. A narrow stretch of bare land extended into the sea. Far off in the distance, across the water, he could see another land.

He returned to his village and told the others. After arguing for days about the risk and the possibilities, a group returned to the land bridge and began walking across. After many days of travel—some of them turned back, a few died from thirst or starvation—they finally reached the far shore. There they saw a miracle: vast herds of huge animals, strange creatures so tame the humans could walk right up to them. They had come to what, much later, you and I would know as America.

This story—of restlessness, of people leaving war or poor resources, or setting out simply to find a better life—continued over the next ten and twenty millennia. Our ancestors risked all, developed new technology, new systems, and found riches they could not have imagined. I have no doubt that some, maybe most, people never attempted the bridge, and stayed behind to survive as best they could. Maybe only a few were bold or crazy enough to walk straight into the unknown. Maybe only a few of these went on to hunt the giant bears and mastodons and tigers on the other side. But these were the people who allowed humanity to expand and thrive, and they became our forebears.

Courage in the face of risks, persistence to overcome long odds, patience to stick to a plan, invention to create new technology, and curiosity about the unknown; even after 50,000 years, these are still the hallmarks of people who move ahead and thrive, pulling the rest of us with them towards better lives.

What if they had never stood up and left their home fires? What if they found the Bering Land Strait, that bridge between continents, and, after walking for a bit, turned back and never tried again? What if humans denied their own native wanderlust, that need to know what lies beyond the far shore? What if they had failed to develop the new technology we needed to fit the new environments? How would we be living today? Would we have evolved at all?

We face the same questions today—questions in the form of a choice. Should we commit to exploring and settling the frontier of space? Or should we play it safe and turn back?

This book argues for the first option. It reveals the urgency of a national space effort, a national (and international) endeavor that can revive the economy and quite literally save humanity. It will show how the financial cost of a robust space program returns riches, including improved security, that outperform almost any other investment. You will see how space does more than any other national venture to stimulate young people to study science and technology and how it helps create the resources for that education. Any one of these benefits is clearly worth the additional cost: one third of one percent of the federal budget.

But this book bears a larger, even more urgent purpose: the survival of the American Dream. I'm a great believer in owning a home and living a comfortable life. But that's only part of the American Dream. The other part is what created America in the first place: impossible goals, the drive to exceed limits, to cross any bridge and build new ones. This dream has us pushing frontiers, settling them, and moving on. Not just figurative frontiers. Literal ones. America is a nation of pioneers. When we stop believing that, we stop being America.

The space program is our next bridge. We need only to build it in order to settle our next frontier and create the next chapter in one of civilization's greatest stories.

That story begins—or, rather, continues—in a region that lies overhead.

INTRODUCTION

1

The Well

We humans, and all the other inhabitants of Earth, live at the bottom of a well which extends a million miles above us. Though we have sent probes that have traveled beyond it, not one human ever has managed to escape. And yet the future of our species depends on doing just that.

Scientists and space engineers call it the Gravity Well. You could also call it our destiny, a continuation of the journey from our origins in Africa to our domination of the skies. This book aims to help us decide what direction we go in the future—which, in turn, affects our present time on Earth. If we begin to make a serious, sustained effort to escape the Gravity Well, we will reap vast rewards long before we reach the top. The Well is a barrier to our future. It's also one of the greatest opportunities ever presented to us.

Which raises the obvious question: What, exactly, is the Gravity Well?

THE SPACE TRICYCLE

To understand the Gravity Well, just for a moment stop thinking of it as a well. Imagine instead a bowl-shaped valley, and a little kid riding a tricycle on the sidewalk at the lowest point. The valley happens to be steepest at the bottom. To climb the hill, the kid has to pedal furiously. Fortunately, there are a few flat places where she can rest without her tricycle going backwards and plummeting back down. She takes her feet off the pedals and breathes. The kid could stay there forever, but she's hungry and wants to get home for a snack. So she pedals again, gaining speed. The hill eases off, becoming less steep, and she keeps pumping. Finally she reaches the top—whew! Down below, in the next valley, she can see her house and her mother, arms akimbo. The kid scoots the trike forward and then lifts her feet as she picks up speed. This kid is flying! While this valley isn't quite as steep as the one she went up, it gets steeper the farther down she travels, and she realizes one big problem with her tricycle: it has no brakes. Gaining some serious velocity, heading straight for her mom, she closes her eyes and… feels a pair of hands grasping her shoulders. She looks up to see her mother right behind her, running while gently pulling back. Mother and daughter slow to a stop. The mother leads her into the house for a snack and a lecture, and the kid looks up and says, "I want to do that again!"

And so you have a rather liberal analogy of gravity wells and the concept of Lagrangian points. The first valley represents the Gravity Well from Earth. The flat spot represents one of Earth's Lagrangian points. The Gravity Well, like the first valley, rises the most steeply at the bottom and eases off as we approach the top. If you look at some maps of space between the Sun and Mars, with Earth in between, the blackness of space is intersected with curved lines that look quite a bit like a topographic map of a wilderness on Earth. Instead of a rising and falling landscape, space has centrifugal and centripetal forces, the pushing and pulling of masses. The "hill" of the Gravity Well is steepest near the bottom because Earth's gravity pulls strongest against the closest objects. The farther up you climb, the less Earth pulls.

Lagrangian points are the "flat" places in space where the pulling and pushing forces are in balance. The orbiting object constantly pushes away from one body, held back by gravity, while the pulling of a second body keeps the object in the same place relative to the two large bodies—the Sun and Earth, in this instance. Each pair of bodies in the Solar System has five

Lagrangian points between them; all five lie in the orbital plane of the smaller body as it orbits the larger one. (Draw a circle on a piece of paper. The orbital plane is that two-dimensional circle on the flat of the paper.) There are Lagrangian points between the Sun and Earth, and another set between Earth and the Moon.

We have placed satellites on Sun-Earth L1 and L2 that observe the Sun, remnants of the Big Bang, and the heavens beyond. The James Webb telescope, Hubble's successor, will be parked at Sun-Earth L2. In the future, Lagrangian points can serve as stable places for bases and resupply depots—and, potentially, large human outposts. The L5 Society, founded in 1975 by fans of visionary scientist Gerard O'Neill, advocates colonies of thousands of people in self-sustaining habitats around the L5 Earth-Moon Lagrangian point, following the same path as the Moon around Earth.

The little girl's tricycle begins picking up speed after she reaches the top of the valley (or the Gravity Well), because she's now entering a second valley. That second valley represents Mars's gravity well. (It turns out that mother and daughter are Martians.) One big challenge of going to Mars is finding ways to slow the craft down to zero miles per hour. In this case, Mom did the job. Crafts going to Mars can deploy parachutes or other drags, but they only provide so much deceleration in the thin Martian atmosphere. A landing module must burn fuel to slow down. And, like our little girl who wants to go back up and down into the next valley, a spacecraft must have enough fuel to leave Mars's gravity well and then to slow down within the Gravity Well of Earth. (When I capitalize "Gravity Well," it means I'm referring to the phenomenon limited to Earth and its satellites, including the Moon.)

In short, space is nothing like a void. It's a terrain of hills and valleys (not to mention actual rocks, dust and planets), with the steep climbs and descents and flat spots made by gravitational and inertial forces.

THE CLIMB

Of course, you can't actually see it, but you feel it every day. It has no physical walls. And yet it is very real; fail to decelerate as you approach the bottom, and your craft will be crushed along with all its occupants. From an engineering standpoint, the Well defines a terrain of space—a steep wall requiring enormous force to climb.

Here at the bottom of the Well, we have already worked wonders. In little more than a century, we built vehicles that left the surface and occupied the space just above it. We built robots and sensing equipment to look down on our planet and allow global communication and navigation. We even set foot on the Moon. More recently, we have begun to create an economy in space, with ever-cheaper satellites and more efficient vehicles. We have sent astronauts to live for months in space.

And yet, that space we are making our own still lies near the bottom of Earth's Gravity Well. Most of today's space economy occupies low Earth orbit, a 1,200-mile-wide band that's a slice of less than one one-thousandth of the Gravity Well.

The International Space Station, more than 200 miles up, also sits low in low Earth orbit. And the Moon? What about the Moon? At 240,000 miles on average, it lies only one fifth of the way to the top of the Well. Imagine if Columbus had reached the New World from Spain and somehow managed to keep sailing all the way to the Moon. His first voyage from

Palos, Spain, to the Bahamas covered about 4,000 miles. From there to the Moon would have been 60 times as far. Think about it. The American astronauts in 1969 traveled 60 times the distance that Columbus sailed to America. And the astronauts had to bring their own oxygen and fuel while pushing against Earth's gravity.

To continue to Mars, we will still have most of the Well to climb—and about 8,750 Columbus voyages of distance to travel, assuming that Mars and Earth are at their closest points in their respective orbits around the Sun, a position they take once every two years.

The good news is, the Gravity Well presents the main obstacle between us and Mars. The steepest part of the Well lies between Earth and the Moon. Once we escape the Well, our vehicles can essentially coast. Mars becomes achievable—assuming we can overcome such technological difficulties as the dangerous radiation that lies beyond the well, and the hazards of landing and living on Mars itself.

But to see the Gravity Well as a mere matter of distance sets the wrong measurement on the task. Escaping the Well is more a matter of speed. To get into low Earth orbit—a path in which the forces of motion balance the pull of gravity—a rocket has to attain a speed of about 17,000 miles per hour. To escape the Gravity Well, a vehicle must go about 25,000 miles per hour— 33 times the speed of sound.

To attain these speeds requires fuel in logarithmic portions. As a rule of thumb, ten pounds of rocket fuel are required to push one pound of equipment or human into low Earth orbit. The fuel itself must be carried too—not just to bring the vehicle to its destination but, if it carries humans, to bring it back as well, accelerating beyond the pull of gravity, then decelerating to keep from crashing back into Earth. And so the Gravity Well constitutes a challenge of speed, ingenuity, and brute force.

Besides presenting a physical challenge—one that requires overcoming the hostile terrain of space—the Gravity Well also represents an ascent of achievement. Toward the bottom lies our airline industry, generating $170 billion a year in revenue. Farther up, in low Earth orbit, or LEO, are more than a thousand satellites, monitoring crops and the weather. Still farther, in middle Earth orbit or MEO, several dozen satellites provide our GPS. At the next level, in geosynchronous orbit or GEO, more than 250 satellites enable our Internet, television, and telephone communications. While governments were first to explore this terrain, it now supports a frontier economy, with basic, low-cost transportation increasingly provided by privately owned companies. This economy is rapidly rising to higher revenues, as entrepreneurs develop more powerful reusable rockets and smaller, more sophisticated satellites to serve a variety of clients and customers on the ground. Still farther up the Well lie the Lagrangian points, with a few more satellites. Farther up lies the Moon, where we Americans left footprints and then chose to pull back.

All of these points—LEO, MEO, GEO, Lagrangian points, the Moon—occupy the bottom portion of the Gravity Well. At each stage, probes and astronauts scout the terrain, and industry eventually follows. As this book will show, our greatest future depends on our escaping the Well. The very effort will reap tremendous, even unimaginable rewards; but only if we choose to make the effort. Future generations will live off a Gravity Well economy that dwarfs the one on Earth, and they will remember our generation as the one that made the first leap. But only if we choose to make it.

OUR EARLY ESCAPE

It took thousands of years for Europeans to develop the technology and the will to explore the New World. Another half a millennium went by before Americans reached the Moon. Little more than a decade from now, humans could physically escape the Gravity Well.

By 2030, we can have a small working colony of humans on the Moon. By that same date, we can be mining ice and precious metals on asteroids. By 2040, we can have a colony on Mars.

By 2050, we can have probes on other, more habitable celestial bodies, more new worlds. All within the lifetime of some readers. We are not done. We have barely begun.

Which raises the question on everyone's mind: Is it worth it? To move beyond the Gravity Well and create those human outposts beyond the Well will take a combined effort by nations, our own government, and private industry, with an annual budget of about $30 billion. This is a great deal of money, even in an economy worth $18 trillion. The amount NASA needs is comparable to someone with an income of $60,000 spending $100 on space. This is still serious money. Besides that, the effort itself constitutes enormous risk to human life.

If we limit our discussion to economics, however, most economists will tell you that a fully funded space program is eminently worth the price. But that misses even more important benefits. The technology required to send people and robots into space creates a huge amount of knowledge that can be used for other things. I call this expanse of knowledge—technology that can be extracted and used by private industry—"knowledge ore." In terms of future wealth, it's the equivalent of gold or oil, only with information.

Let me give you one small example. Intuitive Machines, Inc., recently developed a small device that looks much like a computer mouse. It allows medical technicians to find a person's vein, painlessly drawing blood or inserting an IV in the perfect place. The mouse achieves this by docking with the blood vessel, much as a space vehicle docks with the International Space Station. That's because the software used to invent this mouse was, indeed, adopted from the software that docks space vehicles with the space station. The engineers at Intuitive Machines, many of whom were veterans of NASA's Johnson Space Center, mined the knowledge ore to create something new, profitable, and good for humanity.

The Intuitive IV catheter (let's call it a "phlebotemouse" for fun) is just one of hundreds of inventions created each year from the space knowledge ore. For every dollar spent on space, the knowledge ore brings back as much as $8. And that's not counting the 65,000 jobs created by the space program—which uses private contractors to do most of the work. (Full disclosure: my boss, Kam Ghaffarian, cofounded Intuitive.)

But the rebound does not stop there. NASA stimulated the creation of modern communications networks and weather prediction by proving that satellites were capable of these tasks. The satellite industry generates more than $300 billion in revenue, an amount that continues to grow 5% to 15% a year. Satellite-based weather tracking has an annual economic value of $11 billion. Of course, these industries stand on their own today. NASA continues to build and manage the satellites used by the National Weather Service, the Weather Channel, and AccuWeather; but surely some private entity might be willing to take this on if Congress decides that's no longer the government's job.

Without NASA, though, what will catalyze the next breakthroughs?

Still, I do not pretend that economics will excite every reader into lusting after a robust space program. So let us take a different tack.

WE'RE ALL GOING TO DIE

Researchers from biologists to astrophysicists use complex computer models to predict the eventual extinction of the human race. Many of these models indicate that our species may survive for another century or two—a thousand years, possibly. Beyond that, however, the outlook seems grim. Name your poison:

> Climate change dries up much of the planet, ruins our food crops, makes the oceans rise, and triggers a refugee crisis and global wars.
> Nuclear war or terrorism causes a breakdown of civilization and mass starvation.
> An epidemic decimates the population too fast for a cure.
> An asteroid or comet smashes into the planet, causing a cataclysm of volcanic eruptions, earthquakes, tsunamis, and a pall of ash that destroys almost all crops.
> Cosmic rays from a quasar silently kill all living things on Earth.

Most of us may have forgotten a real-live doomsday scenario that I left off the list: the hole in the ozone layer. During the 1970s, scientists speculated that chlorofluorocarbons, or CFCs, posed a risk to Earth's atmospheric sunscreen, a layer of oxygen molecules that shield the planet from deadly ultraviolet rays. Research confirmed that a hole in the layer had opened up over Antarctica. NASA sent up satellites during the mid-Eighties to measure the extent of the problem. From the vantage point of space, they proved that the hole not only existed, it was expanding rapidly. Eventually, the ozone would disappear altogether. The research was so conclusive that, in 1987, every member of the United Nations signed a protocol to phase out the use of CFCs in aerosol sprays and refrigeration. Scientists now predict that the ozone layer will fully recover by 2070. In short, we are not going to die from hairspray.

But what about the problems that remain? NASA satellites look back at Earth more completely than any other nation's satellites, public or private. We happen to be very good at taking selfies of our own planet. These satellites show the amazingly complex relationships among the atmosphere, the oceans and land ecosystems—all the systems we have to know to correct those factors that tend to make our planet inhabitable to humans.

SAVING THE WORLD IS EDUCATIONAL

Another positive point arises from the threat of human extinction. As a father, and as someone who speaks frequently with young people, I have noticed that smart, ambitious youth tend to fall into two basic categories: those who want to change the world, and those who want to save it. Either ambition can lead to great things, or blow everything up. To channel the ambitions of our best and brightest youth, I believe we need more than movies about space. We need a tangible, crucial mission that saves humanity.

At any rate, we definitely need more people to go into STEM—Science, Technology, Engineering, and Mathematics.

America depends on its technological leadership for security and a healthy economy. Yet we are in danger of losing that very technological leadership. Our university system, as challenged as it is politically and economically these days, continues to be seen as a model by the rest of the world; more than half of students pursuing Ph.D.'s in science and tech in this country are foreign nationals who return to their home countries. Those places used to be occupied mostly by Americans. Inspired by the space program, the number of American math and science Ph.D.'s more than tripled during the Sixties. After the end of Apollo, that number began to decline, and today there is a serious shortfall. America spends over $810 billion a year on education, but it has trouble recruiting its own young citizens into STEM. Meanwhile, other nations are doing all they can to invest in STEM education, with the hope of leading the world. This isn't a bad thing. When we colonize space, we won't do it alone. But if we don't take the lead, who do we want to lead us?

To regain our technological leadership, to continue to explore and discover and invent, to lead the world in saving humanity, and in the meantime to revitalize the economy—we need a government effort as well as a private one.

In other words, we need a bold national space program and more stable and rational funding of NASA. Private industry is building most of the equipment that goes into space, but the leading-edge research and risk-taking have to be spread over our entire society. No single company in its right mind would subject a relatively few stockholders to this risk.

The problem is, NASA can't sustain itself on its current budget. It currently gets $18 billion a year from Congress, an amount that has been shrinking in real dollars for the past two and a half decades. While $18 billion constitutes an impressive amount of money, our spending on Hollywood shows where our priorities currently lie. The 2015 space movie, *The Martian*—the one in which Matt Damon plays an astronaut-botanist—took in more box office receipts on its opening weekend than NASA's daily budget. Our nation has become better at playing pretend than actually boldly going.

To boost that budget to the $30 billion necessary for colonizing Mars and meeting NASA's other, ambitious goals, would constitute an increase of about one third of one percent of the federal budget. To get that money will require a whole different conversation among our political candidates and thought leaders, as well as you and me. We will need to recognize what space does for us as a nation—technologically, educationally, globally, economically. Spiritually. Space, ultimately, is about faith—faith that our curiosity and restlessness, our ability to take some risks, will lead to treasures we cannot imagine.

And yet we see increasing evidence of a loss of faith. We limit ourselves to our comforts and to absolute safety. Our political leaders tell us only how dangerous the world is. We talk about protection. We turn inward, fear the unknown, circle the wagons, and dread the future. Like China in centuries past, we talk about building walls.

To restore our faith in ourselves and our future, we have to put ourselves in the shoes of our forebears, the first Americans who came to the land bridge that stretched through hostile waters toward an unknown land. I imagine they must have held long debates about the risks and rewards of pushing forward. Only those brave souls who took the risks eventually saw the rewards.

Are you, personally, going to be the one who says, "Turn back?"

THE CONTINUING MISSION

America's future continued with the "discovery" of the continent by Europeans. They saw America as a treasure of exploration—and an escape from war, repression and hunger. Our future moved forward with Lewis & Clark, and the settlement of frontiers from coast to coast; with our exploration of the deep seas; and with the systematic effort to put Americans on the Moon.

We think we know the story about the Moon landing. If we think about it at all anymore, we remember—what?—Neil Armstrong taking one long downward step onto the dusty lunar soil. We remember his words about a "giant leap for mankind." We remember the American flag, the footprints, and maybe Alan Shepard swinging a 7-iron, driving the ball forever in the weak gravity. We remember, if we remember at all, that in that moment America reached its apogee in space. After that, it was all downhill: the other landings, the Space Shuttle, the International Space Station, the robots sent to distant planets. Impressive feats, all of them, but not compared to landing on the Moon. That's the story we know; one we ended too abruptly.

This leads to another, painfully obvious question (painful, at least, to me): Why should you care?

This book seeks to answer that question, by first dealing with the most common objections to the space program.

Except for tourism, aren't we pretty much done in space?

After all, the space race is over. We've already set foot on the Moon, a foreign body in space. We've sent probes all the way to Pluto and beyond. We've even put robots on Mars. What else could we possibly do?

In this book, you will see how NASA and its partners are making daily discoveries about the Sun and the Solar System, about the origins of the universe, and about the future of the planet we occupy. What else can we do? We can mine the water that lies beneath the poles of the Moon. We can explore lava tubes on Mars that may be warm enough to sustain life, including human life; build floating stations in Venus's dense atmosphere; break through the ice into the liquid ocean of Jupiter's moon Europa; see if Saturn's moon Titan, remarkably similar in size and structure to Earth, is suitable for settlement; mine asteroids for gold; and reap the unknown treasures on Earth that arise from the inventions required for these missions.

Meanwhile, in this book you'll learn about NASA's current work: how the agency is integral in upgrading the nation's air traffic control system, in predicting the weather in space as well as on Earth, and even in creating the elusive Personal Air Vehicle, the flying car.

The short version is this: We have taken baby steps into space. We're in the position of Spain after Columbus returned from Hispaniola. Spain had a choice. It could have said, "Been there, done that," and saved itself the trouble of exploring this New World. It could have chosen to forfeit in advance the wealth and power that followed.

NASA's current work is essential but not sufficient. The next steps entail moving in. First, though, let's continue with the reasonable objections.

We can't afford it.

Although affordability is the biggest objection to space, it happens to be the easiest to rebut. The answer is not the usual,

"It's a great investment," although space is indeed a superb investment. The government space program actually offers a bigger payoff, in the form of a rich intellectual resource. This book will explore this resource, created from space research and technology. It remains one of the least understood, most hidden of our national treasures.

First, though, to understand this treasure, I want to share with you one of the most important concepts of this book: **Space is more than an act of exploration. It is an act of creation.**

Invention is the twin of discovery. The intellectual public asset created by the American space program generates more wealth than all the timber on our public lands. This is not counting the immediate economic effects of the program itself. In Chapter Six, we'll explore the astonishing range of products and innovations that came out of the creative harvest of this virtual ore of knowledge.

The indisputable value of space's intellectual public property makes a powerful argument for the role of government in catalyzing American entrepreneurship. And that's without mentioning the value of increased international influence, and the demonstrated value of inspiring young people to go into STEM.

But we have other budget priorities.

America has urgent unmet needs: poverty, education, security, infrastructure, health. Back when NASA took up 4% of the federal budget, the "nondiscretionary" part of that budget, comprising Social Security and social welfare programs, were a fraction of the size they are today. Where is the room for space in that budget today?

This question will take the rest of the book to answer. Briefly, though, the answer comes down to this: the federal budget needs a healthy economy to fuel it. Aerospace catalyzes the markets and industries that drive our current economy, including air transportation, aviation, airplane manufacturing, satellites, weather prediction, and a host of high-tech ventures from software to materials technology. The space program lies quite literally at the leading edge of the economy. Fail to maintain that edge, the economy slows, and government tax receipts fall. The space economy drives the American economy.

But why government? If space is such a valuable way to invest, why not leave it up to the free market? Hence the next objection to space...

Private industry can do it better.

In most media portrayals of NASA, rocket ships are built by government rocket scientists. Government astronauts build government space laboratories and government bases on other planets. Government crews on government starships explore distant galaxies.

In reality, private business is involved in virtually every piece of equipment and software of every NASA space mission. The space agency does the design and acts as a general contractor, soliciting bids, structuring, integrating, and guiding the mission. Private industry does much of the heavy lifting—literally. The fuel in booster rockets comes from private chemical companies. The rocket engines themselves are built by the likes of Pratt & Whitney, AeroJet Rocketdyne, Blue Origin, and Orbital, as well as the nations of Russia and Ukraine. Federal employees comprise just a fraction of the total NASA mission workforce.

So why have government involved in the first place? Two primary reasons: initial investment and risk. NASA opens frontiers deemed too risky for private investors and too distant in time for profitable operations. Take, for example, the new companies springing up to create space rockets. They are leveraging and innovating with decades worth of government developed technology. Plus, the government indemnifies these companies, providing a critical insurance policy against lawsuits in the event of failure. Or consider the prospect of mining a nearby asteroid filled with platinum (a definite possibility). Several startups have worked up business plans to exploit the vast riches. But only NASA has the wherewithal and technical prowess, and can take on the risk, of technological development and exploration. In every aspect of space discovery and exploration, the government, through NASA, fronts the money necessary to prove a concept before it can be believably profitable, in much the way the National Institutes of Health sponsor research on diseases that eventually result in profitable drugs. Government is a catalyst for the creation of new markets and industry that most people cannot imagine at the time. Through government, society spreads the risk of failure through an entire nation rather than one company, with the confidence that the nation as a whole will profit.

What's more, the intellectual property created by the space program is public property. Its public nature allows an unimaginably broad range of industries to tap into it. No one company or industry owns it. That's why space docking software can be employed in docking with patients' arteries. The intellectual space resource is a broad, richly diverse vein of ore. Creative minds will discover wildly different ways of using it. The most entrepreneurial among them will benefit directly. But so does the nation as a whole. If you ask serious economists, they will tell you that every region and a host of industries—from medical to energy to sports to household appliances—benefit directly from the intellectual discoveries of the space program.

Without a bold public space program, that resource will age and eventually become useless. Without the government spreading the risk throughout society, new discoveries will not arrive, less knowledge will get created, and new markets and industries will fail to emerge. Space attempts will become more modest. What once was a rich resource will become sparse. Intellectual property will lie in the domain of private companies guarded by fierce lawyers. We're already going down this path. The economy will plod along, no doubt. But it will be a more static, flat economy; much like other nations, those that are past their prime and receding in influence globally. America will be that much less rich, and there will be fewer entrepreneurs getting wealthy from one of the nation's greatest resources.

Still, the economy alone fails to excite people about space, understandably. President Kennedy didn't announce that we were going to put people on the Moon within a decade simply to stimulate the economy. Though indeed the vigorous space program catalyzed a new commercial satellite industry for global communications, a weather forecasting industry, microelectronics breakthroughs that enabled the computer and internet industries, dozens of medical and materials discoveries—new companies and products we now take for granted. The direct benefits arising from Kennedy's initial challenge now drive our economy with hundreds of billions of dollars in annual activity, none of which existed before he took office.

But economic benefits were not Kennedy's reason. For him, the reason for space was international security and American influence. That is why, with the end of the Cold War and our conquest of the Moon, it's hard for many of us to see why we should strive to truly be a spacefaring nation.

Space has to have a reason.

Other countries should bear the burden.

More nations are heading into space—a dozen of them as I am writing this, and growing. An orbital satellite has become a must-have item, along with an airline, for any self-respecting nation. Meanwhile, China, India, and, possibly, Iran are upping the ante in space.

This book will briefly explore the implications of letting competitors like China and Russia take the lead in space. Every military general will tell you that the essential battle tactic is to gain the higher ground. Space is the ultimate high ground. It's also where we have taken our communications and the devices we use to locate ourselves. It would be foolish to give up this territory to powers that someday may want to do America harm or who could beat us to colonies on the Moon or Mars, with political systems that lack our commitment to freedom.

But there's a still more noble reason for space—a peaceful reason. Despite the growing efforts by the other powers, none of them has the means or the technical ability to meet our mutual ambitions in space. The future (perhaps the very near future) of humankind depends on our ability to explore and colonize other celestial bodies and to understand the forces in play between them. This must not be the sole endeavor of a single nation; nor can it be left to a race between rivals. To achieve our greatest goals in space requires a human-wide effort. That means cooperation between nations, even rivalrous ones.

And here's the good news: We've already been cooperating for more than three decades. Fifteen nations currently support the International Space Station. Even while tensions were rising over Russia's domination of eastern Ukraine, Russians and Americans were conducting science experiments, living and eating, performing maintenance work, conducting mutual ground-based operations, and communicating aboard the Space Station. Our cooperation in space provides a counter-narrative to the story presented by the news media and government spin doctors. While hostility grows between the two nations, scientists and engineers are gaining new knowledge in the most hostile of environments, learning the effect of space on humans, how to build reliable systems that work in a gravity-less vacuum with dramatically changing temperatures. Neither nation is "winning." Neither is losing. Humanity is gaining.

This is more than a story of international cooperation. It's also a story of American leadership in space. The new chapter of space exploration will not be about Americans going alone to other planets. But it can be a story about America leading the way. If NASA is given the means and the direction, we can guide the multinational effort into space. We can prove that American leadership means more than winning on the battlefield, more than defending ourselves against threats to our security. To regain the deepest respect of the rest of the world, we have to lead the rest of the world, not just protect against it.

In short, the world once respected America for acting more out of courage than fear. The challenge of the Gravity Well will let us choose courage once again.

No, we're not done. We have barely begun to break out of our Well.

STORY

2

What's Still Out There

NASA is not just astronauts. Nor does it exist only for the physical exploration of space. America's space agency isn't even just about space. Its major initiatives comprise astrophysics, the purest science of the cosmos; study of our Solar System; aeronautics, focusing on improvements to aviation technology; Earth science; examination of the Sun; space technology, which finds new ways to conduct robotic and human missions; and, finally, the human exploration program.

In this chapter, we'll go on a quick flyover. First, though, let's land on one of these branches and examine a single effort. You need to feel what it's like to be one of NASA's non-astronaut explorers. We will do that by replaying an effort I worked on not so long ago, working with a diverse team of some of the most brilliant minds in America. Our mission: to create a terrain map of space—not of planetary bodies but of the clashing forces between them.

But wait. How is mapmaking a kind of exploration? Isn't it merely a prerequisite to exploring? To understand what I mean, you need to know an essential concept of the ancient Greeks and Romans. They used this concept to develop two of the most powerful, creative civilizations in human history.

The word for this concept is *inventio*. It means both invention and discovery. The ancients believed the two are inseparable. Nothing gets created wholly anew; it was an ancient Greek who wrote, "There is nothing new under the Sun." Invention meant discovering the available means to solve a problem or do something new.

Engineers understand this concept implicitly. The team that invented the iPhone did not invent the digital music player, or GPS, or email, or the Internet browser or radio communication. They discovered a new purpose for each, and married them into an intuitive machine for Apple's smartphone. Of course, a great deal of technical wizardry was required to make all these existing inventions play nice together; but the result was a pure act of *inventio*: invention and discovery in one creative act.

You could say that *inventio* is the perfect description of NASA as well. It invents in order to discover; and the inventions themselves are an act of discovery. While most NASA engineers wince when you call them "rocket scientists," in a way the label is appropriate. The agency comprises deeply practical people with a passion for discovery. They invent by discovering, and discover through invention.

Back to the mapmaking. In 2012, when I served as the Engineering Director at Langley Research Center, NASA assigned a group of us the task of determining where to park a new outpost in space, one that would provide a waypoint between Earth and the Moon—an essential step to colonizing the Moon and then Mars. For four months, I commuted four days a week between my home in Virginia and the Johnson Space Center in Houston. I joined eight other engineers, mathematicians, and physicists in a meeting room dominated by a projection screen and a whiteboard. My fellow team members included experts in navigation, communications, rocketry, physics, and a bevy of other disciplines, including one astronaut. Taking turns with markers, we scribbled schematics, mathematical formulas, and mechanical drawings. We argued over politics, the sexiness of our locations, and the expense. But mostly we focused on the task at hand: to determine the ideal mission and point in deep space for the new outpost.

This, again, constituted an exercise in pure *inventio*. The theory behind Lagrangian points had been around since the mid-1700s, when they were conceived by an ingenious mathematician named Joseph-Louis Lagrange. Born in Italy, he moved to Berlin to become the director of mathematics at the prestigious Prussian Academy of Sciences. He ended up in revolutionary France, where he became instrumental in development of the metric system. Lagrange helped invent probability theory and advanced number theory. He corrected Newton's theory of sound, pioneered the study of work efficiency; and his calculus equations led to the greatest advances in physics in the nineteenth century. Oh, and Napoleon, a huge fan (the feeling was not mutual), appointed Lagrange a senator in the French assembly. Engineers revere him for the math he brought to celestial mechanics.

In 1764, using a set of differential equations to calculate the gravitational dance of the Sun, Earth, and Moon, Lagrange explained why the Moon has one side always turned away from us. Eight years later, he helped solve one of the great mysteries of outer space: the "general three-body problem." Take Earth, Moon, and Sun. You know the initial position, the mass, and the velocity of each. You know Newton's laws of motion and universal gravitation. Now explain how all these forces act upon each other. Lagrange's formulas helped form the math that determined the Apollo flights to the Moon.

But not until the 1980s did NASA begin exploring the deeper possibilities of Lagrange's math. Then the agency's scientists and engineers began imagining the forces of mass, inertia and gravity as a terrain. Our job was to figure out which point in that space landscape was ideal for the new station. We had two good choices between Earth and the Moon: the Lagrangian points named L1 and L2. We had to come up with a plan for getting the station there safely, and parking it—stopping it—in exactly the right spot without wasting fuel. In addition, our job was to help create a railroad grade through space.

Let me explain in the simplest Lagrangian terms. The Apollo program was the equivalent of the Lewis & Clark expedition, a set of explorers who took the straightest possible path (as much as any path that involves multiple orbits around planetary bodies can be straight), over hills and mountains, using whatever energy it took to go from point A to point B. The astronauts went to the Moon, taking everything they needed with them, and then they returned home, all in about a week. My team dealt with the problem of shuttling equipment, supplies, and teams of people back and forth, not once but many times. Following the same path as Apollo would require enormous expenditures of fuel. It would be like trying to build a railroad simply by following Lewis & Clark, with cliffs and canyons along the way. Instead, we needed to discover the best grade, a "flat" course with the most neutral gravitational forces. This meant going a much longer distance to avoid the rough places.

We decided that L1, located closer to Earth than the Moon, was too easy. Getting there meant barely leaving our terrestrial neighborhood. L2, on the other hand, presented a challenge. It lies closer to the Moon than to Earth, and it would stretch our technological prowess to the fullest.

Four days out of every week, we met in front of the whiteboard and hashed out choices and plans. Some weeks, industry teams would join us to show their ideas. Then we went back to our homes and laboratories with individual assignments, enlisting small teams of brilliant NASA and university engineers in a host of disciplines to put meat on the bones of our concept. We presented our findings each following week at the Johnson

Space Center. Our end result: a detailed plan consisting of designs, analysis, drawings, trajectories and cost estimates to park a space station at L2 with a direct view of deep space and the far side of the Moon.

Seven months later, NASA scrapped the mission concept. Agency leaders chose not to send a station to L2. Instead, NASA would focus on an entirely different task: to land a robot on an asteroid and bring back a sample. When we heard the news, we were understandably disappointed. But our work wasn't in vain.

Seriously: our work was not in vain. There's a reason I told this story, and veered off to tell you about the ancient Greeks' and Romans' notion of *inventio*, along with the whole backstory of Joseph-Louis Lagrange. Without a single one of us making the leap into space, our team had explored it. We did it with math, powerful computers, and a whiteboard. Our calculations and plans can also be used to get to the asteroid—and to Mars for that matter—and they will help catalyze other missions. Essentially, we were cartographers, and that is no small thing. Christopher Columbus's mission to the New World was an act of cartography: his initial assignment by the king and queen of Spain was to map a course from Europe to Asia. Getting to China wasn't really the point; not to mention the fact that there happened to be a continent in the way. The point of Columbus's mission was a map, one that later voyagers could follow. The team I worked with at NASA did the cartography in front of a whiteboard and in the computers of our teams around the country.

You don't see this sort of work in the news media, understandably. Just about every week the news bears the latest successful NASA missions, complete with colorful photos. Meanwhile, some of the best minds in the agency are working on tools and methods that never get off the ground—not, at least, in ways the teams intend. No private company would put up with this kind of R&D of constant apparent dead-ending. Spread out over a hundred million taxpayers, however, the risk becomes positively palatable. And ultimately it produces benefits for the greater good. The missions may never leave the ground, but they live on as acts of invention and discovery. The tools and methods get used by other missions and worked into ever more sophisticated tools and methods. Our labors on L2 are getting incorporated into the asteroid and other missions. Multiply this effect among the many major programs at NASA, and among the dozens of current missions, and you have a vast creative engine of *inventio*.

There's just one devastating problem: the asteroid mission has not received the go-ahead, either. The dead-ends keep multiplying.

Still, teams continue to work on missions that may or may not get off the ground, creating a vast wealth of data and technology across all four major branches of NASA (aeronautics, space technology, science and human exploration; science includes astrophysics, Earth science, heliophysics, and planetary science). This book is not meant to be a comprehensive survey of the agency; but it helps to see how much NASA permeates and drives the nation's economy, science, and technical R&D. Let's start with the science.

PURE SCIENCE: ASTROPHYSICS

Astrophysics seeks to explore everything from the outer fringes of the universe to the beginning of time, all while searching for

Earth-like planets. The best-known program in this branch—the Hubble Telescope with its eight-foot mirror and instruments measuring visible, infrared, and ultraviolet light—has punched above its considerable weight. Operating in low Earth orbit about 350 miles above the distorting atmosphere, Hubble has produced more data and photographs (not to mention images of gas clouds and star clusters that amount to art) than the telescope's creators could have imagined. Scientists have used its observations to measure the universe's rate of expansion.

Eventually, Hubble came to the end of its productive life, and its control systems began to fail. To fix the problems and extend the telescope's lifespan would mean risking the lives of astronauts. The 2003 disaster with the *Columbia* Space Shuttle, in which all seven crew members died on re-entry, made us all the more cautious about the risk. We decided to shut Hubble down. And then came a huge outcry, not just among scientists but from the general public. NASA Chief Sean O'Keefe began receiving 400 emails a day from "Hubble Huggers"; ABC news reported that "thousands" of schoolchildren wrote letters. Senator Barbara Mikulski led the opposition in Congress, and Representative Mark Udall introduced legislation requiring an independent panel of experts to look into the matter. The National Academy of Sciences piled on with criticism of the decision, leading to the resignation of an important NASA administrator. The agency soon relented and in 2009 sent up a crew to repair the equipment and replace its systems—including a UV instrument 35 times as sensitive as its predecessor. The astronauts completed the job in two spacewalks, and to this day Hubble is going strong. Some experts think the telescope could be bringing back images of deep space into 2030 and maybe beyond. (Note the critical role of astronauts and robots working together for the biggest bang for the buck.) The telescope's fan base also remains strong; its Facebook page has more than two million likes, and more than 1.5 million are following it on Instagram.

The Hubble is a tough act to follow, but the James Webb Space Telescope, due to launch in October 2018, will seek to outdo its predecessor. Parked at Lagrangian point Sun-Earth L2, it will have some impressive equipment onboard, including a 21-foot mirror and unparalleled resolution, allowing a visual glimpse of the first stars in the universe as well as star-forming clusters. The Webb, in short, will show us our own beginnings. Imagine a video camera that allows us to witness the signing of the Declaration of Independence or Caesar's assassination— only we're talking not mere humanity but the beginning of time itself.

Now imagine being on one of the dozens of teams assigned to this task. They began work in 1996, collaborating with scientists, engineers, administrators and diplomats from 17 countries. The plan was to launch in 2011, but cost overruns and technical problems kept setting the date back. After NASA spent $3 billion on the project, the House of Representatives voted to shut Webb down. The Senate brought it back to life.

While this story may sound a bit too inside-the-Beltway, it's one that permeates all of NASA's programs. Teams of rocket scientists work with passion on projects that go just so far before wobbling perilously on the vagaries of Congressional waffling. In this case, there is hope that the Webb will end up in space, parked a million miles from Earth and rewriting the history of the universe. We'll never want to shut it down.

THE VENUSIAN CAUTION: PLANETARY SCIENCE

The Solar System operates as a vast climate experiment, a story whose moral applies directly to us Earthlings. NASA's Planetary Science Directorate sends up probes and robots to our neighbors, retrieving data that reveal the life history of planets.

We discovered, for example, that Venus used to be a pleasant, watery, habitable planet much like ours. Gradually, over millions of years, the Venusian atmosphere became increasingly acidic, with more carbon dioxide. The CO_2 trapped the Sun's rays, causing the planet to warm dramatically. The Martian atmosphere, it turns out, also used to be much more like Earth's. Mars, too, built up its CO_2 levels. We're not certain what happened with either planet. What caused the rise in the Greenhouse Effect? Finding out could help us discover solutions for our own planet. Our use of fossil fuels is accelerating the process that took our neighbors millions of years.

Ironically, our studies of the planets in our Solar System could also help pave the way to escape our own planet, or at least diversify our inhabited real estate. The Solar System Science Division produces information that directly feeds into future missions, including human missions.

Meanwhile, probes and robots have proven popular with the public—especially the four Mars Rovers that have been exploring the Red Planet since 1997. The *Opportunity* Rover holds the distance record of 25 miles on the surface. It discovered that Mars used to be warm and wet.

The *Opportunity* and *Spirit* Rovers, sent up in 2003 with a designed life expectancy of three months, kept roving for eight years until they finally began showing their age. As with Hubble,

NASA decided to end the project. Once again, public outcry led Congress to continue rover operations. Eventually, years beyond its design life, *Spirit* shut down. In 2011, NASA launched the nuclear powered *Curiosity* rover, a vehicle the size of a small car, to the surface of Mars. Today it and *Opportunity* are still operating. We are learning more and more about the climate changes that made Mars the cold, dry and CO_2-saturated planet it is today.

AERONAUTICS: THE ELUSIVE FLYING CAR

We tend to think of NASA as a space agency, period. But that first "A" for aeronautics in the acronym says a lot about the agency's evolution. Its predecessor, the National Advisory Committee for Aeronautics (NACA), invented superchargers for high-altitude bombers, pioneered in wing design during World War II, and provided critical research into breaking the sound barrier. (One of its major laboratories, Langley, was my home base for 28 years.)

President Eisenhower replaced NACA with the National Aeronautics and Space Administration (NASA) in 1958, as part of an effort to launch a satellite into space. Aeronautics remains deep in the agency's DNA, and it accounts for an essential set of programs.

Aeronautics is the science and technology of atmospheric flight—in short, anything made by humans to fly, from wings to propulsion to the forces that make the combination take to the air. Aviation contributes $1.5 trillion—more than 5%—to the national economy, and it's one of the few American industries with a positive balance of trade, with a million manufacturing jobs.

Besides providing the brains behind flight, NASA's aeronautics programs work to keep commercial flights running safely with minimal delays. The challenge is to reform the nation's air traffic control system—a network that uses radar technology developed early in the last century—and add satellite navigation. Besides helping coordinate planes in the sky, the new system will allow the more efficient use of runways and airports, a win for both passengers and our carbon footprint. NASA is also developing next-level technology to automate flight, through development of what we engineers call "autonomous cyber-physical systems"—in other words, flying and space vehicles controlled by computers.

One outcome of this work: the autonomous flying car. You may have seen some of the new flying cars—personal air vehicles, or PAVs—on the Internet. Equipped with folding wings, they can drive on the highway and then, controlled by a licensed pilot, take off from any standard runway.

There are two reasons why we have not seen flying cars in our early-adopter neighbor's driveway: licenses and runways. The current flying cars are really just highway-enabled airplanes. To make a true flying car requires vertical takeoff and landing, or VTOL. Instead of runways with a minimum length of 1,700 feet, you could have designated parking areas barely larger than the car itself. That would eliminate the runway challenge. Then there's the license part: it takes skill to fly a small plane, and even greater skill to pilot a VTOL vehicle. The answer is to create a *self-driving* vertical takeoff car—a Google car of the skies. This flying car will use a combination of laser sensors, cameras, and satellite-based GPS to sense its location and avoid obstacles. Powerful computer calculations will help it anticipate and avoid bad weather or human error.

NASA has been working on the autonomous part of the PAV, and the technology is advancing rapidly—possibly faster than the vertical part. Getting a vehicle to take off and land vertically, safely, with efficient use of fuel, under a variety of weather conditions, in crowded cities or suburbs, ideally without deafening the neighbors, presents a big challenge. But I wouldn't be surprised if the first street-legal and sky-legal PAVs came on the market by 2025. I'm guessing the makers will include older aviation companies as well as a startup or two. We'll all see the PAV as a triumph of capitalism; and it will be, thanks in part to NASA's catalyzing technology.

EARTH SCIENCE: FROM WEATHER FORECAST TO CLIMATE FORECAST

Anyone wondering what NASA has to do with Earth science would do well to take another look at the classic photo, *Earthrise*, showing our planet from space. The agency's work lets us step back from Earth and see it as a whole, in all its complexity. NASA satellites and instruments have been critical in studying Earth's climate, including the factors, or "forcings," that cause change: solar radiation, greenhouse gas, and the tiny particles emitted from soil and smokestacks. NASA also studies the "feedbacks" that increase or diminish warming. Take ice, for instance, which reflects the Sun, bouncing the rays back into space. When ocean ice melts, the surface hue changes from light to dark. The Sun's rays bounce off white ice; dark water absorbs the rays, warming the ocean and melting still more ice. This effect, called ice-albedo feedback, contributes greatly to global warming.

Other feedback loops get intense NASA study, including clouds (like ice, they reflect the Sun's rays); forests (they remove about half of carbon dioxide emissions each year), and precipitation (it helps plant growth, which absorbs greenhouse gas).

But NASA does not limit itself to examining the contributors to climate change. The agency also creates tools to predict that change, pushing the time horizon of accurate weather forecasts deeper into the future.

To understand what I'm talking about, it helps to know the terms meteorologists use for "weather" and "climate." Climate is weather over time. Any forecast under two weeks predicts the weather. A prediction that projects beyond two weeks qualifies as a climate forecast. For example, today we can predict El Niño cycles lasting months or even years. These predictions count as climate forecasts.

Weather predictions have been pushed ahead dramatically since the 1960s, when an accurate forecast maxed out at three hours. Today, meteorologists can predict weather with equal accuracy up to three days. As a result, the weather report has gained tremendous value—totaling about $11 billion in this country alone. Still this amount is a fraction of what the forecast could be worth if it stretched into the realm of climate. Weather, climate, and natural, weather-like hazards such as volcanic eruptions affect $3 trillion, or one-sixth of the national economy, every year. Billions of dollars get lost in ships waiting in harbors for an uncertain harvest. The energy industry can only guess how much natural gas a region will need in any particular winter. The implications throughout the economy are almost limitless.

New satellites and computer technology promise to push the three-day accurate prediction further toward climate prediction. But Congress cut the federal budget for prediction models just as the Europeans were making a serious, well-funded effort. A decade ago, the European Union laid out the social benefits from long-term forecasts and spent the money necessary. The resulting models showed their superiority before Hurricane Sandy in 2012. The European Medium-Range Weather Forecasts predicted Sandy's path toward the East Coast, beating the American models by a matter of days. The American models predicted that the hurricane would veer off into the Atlantic. Homeowners and the authorities thus failed to take steps to reduce the harm. As the Europeans predicted, Sandy slammed into the Northeast, leaving eight million people without power and causing $50 billion in damage. Improving hurricane track forecasts is worth serious money. The economic cost to local and state governments of evacuating just one mile of coast before a storm hits totals more than $1 million. If more precise predictions could reduce the expected landfall swath, millions would be saved with every storm. In one year, these savings could pay for the satellite and the computer model developments needed to improve the forecasts.

NASA is at work to help improve the satellite imaging and computer models, but Congressional funding continues to lag behind the Europeans' efforts. Meanwhile, the Chinese are launching a fleet of Earth science satellites, adapting American technology with newer equipment. If things continue the way they are, we'll be renting our weather future from the competition.

HELIOPHYSICS: THE SPACE WEATHER REPORT

The Sun constitutes the largest atomic physics experiment in the Solar System, with a series of explosions and release of radiation that determine the very existence of life on Earth, and could just as easily destroy it. Every time the Sun burps (a "coronal mass ejection" in scientific terms), it sends out a spray of particles, a deadly rain that scientists call "space weather." The magnetic field surrounding Earth redirects these charged particles so they pass by, keeping the killing rain of rays from bathing our planet.

The Moon lacks this robust magnetic field. As a result, a solar flare could sicken or kill humans. The solar weather report can constitute a lifesaving prediction, and we can't make further advances in space without it.

But the biggest benefit of heliophysics, the study of the Sun, has to do with our climate on Earth. The Sun's complex emission of radiation and particles interacts with Earth's magnetosphere as well as the Van Allen radiation belts, an ear-shaped pair of loops of charged particles that surround Earth. This interaction can distort worldwide communications, disable satellites, bring down power grids and, most important, heat or cool the planet. NASA, in collaboration with NOAA and other nations' space agencies, operates more than a dozen satellites and instruments to find the most compatible ways to live in a neighborhood with a star.

HUMAN SPACE PROGRAM: PEOPLE, ROCKETS, SETTLEMENTS

Granted that we have much more work to do, studying space and our planet, making human flight easier, predicting the climate. Each of these efforts costs less than the price of sending humans into space. We've set foot on the Moon. We've occupied space stations. What other feat would be sexy enough to warrant the cost?

We'll deal with the economics in a later chapter. But, since we're conducting our high-altitude flyover of NASA, let's look at where the human-in-space element fits.

The human space program consists of several basic parts:

A large rocket capable of lifting people, their equipment, and life support beyond the Gravity Well of Earth, the Moon, and other planets.

A manned spacecraft capable of reaching a destination safely and then returning unharmed all the way back through Earth's atmosphere to the surface.

Space stations or outposts.

Communications and navigation networks that allow astronauts to speak with Earth from deep space.

Right now we have one outpost, the International Space Station, orbiting Earth at a cost of $3 billion a year. To achieve the goal of occupying Mars, we will need an optimal vehicle that can move between Earth and the station. We may build an additional outpost in orbit around the Moon, and we'll need a vehicle for this trip. Then we'll need another vehicle to make the journey between the Moon (or its orbital outpost) and Mars.

Why the outposts? Because half the energy needed to send a vehicle from Earth to another body gets spent on escaping the Gravity Well. Once in orbit, the costs of going somewhere else decline exponentially. While the Moon has only one-sixth Earth's gravity, the fuel required to escape the Moon's gravity well must be carried from Earth. Eliminating the need for that fuel, by establishing an outpost in lunar orbit, would save that cost as well.

To bring those humans and their equipment into space, NASA is building its biggest rocket ever, the space launch system or SLS. The first version will push 77 tons into orbit. Eventually, the plan is to build a rocket that can carry 143 tons into space—10% more thrust than the Saturn V.

We have grown so used to seeing rocket launches that it's easy to forget the vast technical difficulty of sending up a vertical bomb in a controlled, directed, sustained explosion. Elon Musk, the famous entrepreneurial founder of SpaceX, has learned that lesson the hard way. Fifteen years ago, he liked to dismiss NASA's slow-paced bureaucracy. He promised to be shuttling tourists into space by now. Though it has taken longer than he had imagined, so far he's achieved 29 launches, with smaller rockets, all unmanned, than those of other companies or NASA's SLS.

This is not to disparage his efforts. Private space programs help advance the technology. But don't be fooled by them: Each program, including Musk's, could not get off the ground without NASA's knowledge ore—including operations, technology, and systems management knowhow. As you will see in Chapter Five, the best entrepreneurship relies on the R&D risk taken by the public space program. The right formula for space includes a powerful public catalyst focused on enabling space entrepreneurs. And keep in mind that almost every physical element of a public rocket launch—every engine, every part, every piece of solid-state circuitry, even the memory foam in the astronauts' seats—gets built by private companies.

Yes, there's a healthy rivalry. But mostly it's a partnership. As the next chapter shows, the proof lies in our nation's history.

3

America in Space

Samuel P. Langley was no stranger to the waters of the Potomac. The Boston-born astronomer and aviator had used the river as a launch pad for his homemade flying machines, the latest of which was large enough to accommodate a man. But on a raw December morning in 1903, the Potomac appeared less like a landing cushion than a watery grave. This was Langley's do-or-die moment. Eight years prior, his work had caught the eye of President William McKinley. A $50,000 research grant from the War Department—worth more than $1 million today—gave Langley the resources to travel to Europe, conduct engineering research, and build bigger flying machines, which were called *aerodromes*. It was an inventor's dream—being bankrolled with taxpayer dollars—until it came time to launch the first man-sized aerodrome. Piloted by Langley's assistant, Charles Manley, the aerodrome crashed into the Potomac. Langley and his team managed to fish the aircraft from the water and ready it for a second launch attempt. Langley could only watch as the aerodrome hovered above the water before plummeting back into the depths.

Fifty thousand dollars down the river.

It could have ended there, on the Potomac. This would be Langley's last attempt at flight. But for the United States government, it was a baby step towards catalyzing the study of aeronautics. The remarkable thing about the aerodrome failure was not the apparent waste of taxpayer dollars. In fact, the money wasn't wasted at all. Engineers learned that, before reliable airplanes could be developed, we needed a systematic understanding of the physics of airflow and structures, as well as the interaction between air flow and the natural vibrations of the parts. There's also a larger lesson. The failure didn't stop us. America went on to explore the problems that caused the aerodrome failure and learned how to control air flow and vibrations. A crash turned into an enormously valuable lesson, rather than a sign to give up.

A BABY INDUSTRY

Just a few weeks after Langley grounded his dreams, the Wright brothers successfully flew and landed the world's first functioning airplane without any federal subsidies. It was a historic achievement, but hardly an isolated one, for aeronautics—as a field of research and engineering—was taking off in Europe. The motivation behind this multinational flight fever was a grim one: the inevitability of war. By 1914, Britain, Sweden, and Russia had established agencies devoted to aeronautics research. France was developing its own airplanes that the country hoped to use in battle. Germany's efforts to launch massive zeppelin airships had yielded a fleet of lighter-than-air bombers that were enormous and horrific to behold. World War I was about to begin, and for the first time in history, much of the war would be fought from the skies, the new military high ground. In this respect, America had fallen behind.

The Wright brothers, in short, made a glorious invention. Now America needed an industry, one that required better understanding of flight science, including aerodynamics, engines, and control systems.

Faced with Europe's advances in flight science, Smithsonian Institution Secretary Charles D. Walcott proposed that the government set up its own aeronautics advisory committee. Its purpose: to direct and prioritize all scientific investigations concerning flight problems and their practical solutions. Walcott won the support of key government players, including a young assistant secretary of the Navy named Franklin Delano Roosevelt. The legislation that would establish Walcott's proposed committee took the form of a rider attached to a naval appropriations bill. President Woodrow Wilson signed the bill into law on March 3, 1915, on the last day of the sixty-third Congress. The resulting Advisory Committee for Aeronautics became known for its acronym, NACA. Known, that is, to a relatively few aeronautics insiders. At the time, few citizens took notice. Headed by Brigadier General George Scriven, the inaugural incarnation of NACA comprised only a dozen men who worked without pay. Meetings took place only a few times each year.

That changed in 1917, when the U.S. Congress belatedly declared war on Germany. Suddenly, NACA found itself flush with money, staffers, and assignments. A special laboratory in coastal Virginia, named after none other than Samuel P. Langley, was built for exclusive NACA use. During the war and for several decades afterward, NACA personnel built wind tunnels to simulate air density changes. They cut dozens of aircraft wing shapes to test indoors and out, and developed cowlings to reduce the drag of an aircraft while funneling cool air to the radial engine. This research produced a range of military aircraft prototypes, many of which were actually shipped to Europe and used for aerial assault and transportation. Rather than keeping their research locked away for cabinet-approved eyes only, NACA released many reports to the American public, disseminating unparalleled volumes of information to the aviation community and setting the foundation for an industry that would reshape the U.S. economy and way of life long after the war.

The work that NACA did in the Twenties and Thirties enabled aircraft makers like General Dynamics, Boeing, and

Douglas, along with airlines like Pan Am and TWA, to build successful industries. Without NACA's airfoils, cowlings, and experimental flight vehicles, the cost to private investors would have been prohibitive. What's more, planes would not have been safe enough for civil flight. Then, when World War II came around, America had a technological lead time that allowed it to achieve air superiority by the end of the war.

AEROSPACE GETS DELIVERED

But we're getting ahead of ourselves. Once the First World War had wrapped up, NACA began allowing military and commercial clients to use their laboratories on a permitted basis. A nascent airline industry had already emerged. The first commercial flight had actually taken place before the war, on New Year's Day, 1914—a hop between St. Petersburg and Tampa, Florida, lasting all of 23 minutes and costing its single passenger, Abram Pheil, $400. Commercial airlines began selling short-distance flights to well-heeled customers, and companies began building airplanes. But the investment costs were large, and one crash would spell bankruptcy. Even with the technological boost from NACA, an additional stimulus was needed. The industry needed a market, and the market couldn't start up without a robust industry. The egg needed a chicken.

Salvation came in 1925, in the form of a mailing envelope. Most government agencies had treated the arrival of the airplane with curiosity and caution. But there was a notable exception: the Post Office Department. As early as 1911, the Post Office had begun authorizing experimental mail-carry flights at fairs and aviation festivals. The prospect of transporting a letter from origin to destination many times faster than a train was irresistible. The experiments grew in scale, coming to a head on February 22, 1921, when the department used a series of private aircraft to transport 16,000 letters from New York to San Francisco. Flying day and night, they beat the best possible railroad time by 75 hours. Impressed, Congress appropriated $1,250,000 of funding for expanded airmail service. In 1925 it passed the Kelly Mail Act, which authorized the Postmaster General to contract with private industry for airmail service.

The story illustrates an essential aspect of the government's role in developing technology. Relatively little tax money goes toward actually building equipment like airplanes. Langley's aerodrome is an exception to the rule, as well as a cautionary tale. Most of the time, the government doesn't build. It buys. By creating a market for the new aviation industry, Congress took the economy airborne. It funded an experiment to prove whether planes were practical for more than warfare. As a result, aviation would boost the American economy and help define the first American century.

Meanwhile, nations across the Atlantic, led by Germany, were spending far greater amounts on science and technology— more than $2 billion in today's dollars just on the V-2 rocket, which began terrorizing London in the fall of 1944.

Once again, America found itself falling behind in technology. The war brought in a fresh infusion of money to NACA, which built two new laboratories, expanded its aeronautical research, and supplied the air industry with blueprints for bigger, more capable war planes. Still, basic science research lagged. It took a visionary government official, and years of effort, to turn America into a science leader. Appropriately, the man behind the vision actually coined the term for the cause: "basic research."

ROCKET-POWERED PLOUGHSHARES

Vannevar Bush first joined NACA in 1938, but his experience as a liaison between military personnel and civilian scientists spanned back to World War I. In his youth he had helped the National Research Council develop equipment that could detect submarines through disturbances in the magnetic field. This endeavor, mining the study of ocean currents for a new piece of wartime technology, put Bush in an excellent position to champion the practical benefits—military, economic, and social—of research. His experience gave him an appreciation of science's long timelines. He used the term "basic research" to describe scientific efforts with both fundamental and applied goals. As the United States and its allies inched closer to victory in World War II, Bush found an advocate for his cause in President Roosevelt. The two had developed a working relationship during the war; Bush had been instrumental in persuading Roosevelt to sign off on the Manhattan Project. Hearing Bush outline the process of bringing researchers and engineers together to build the world's first atom bomb turned out to be a light bulb moment for Roosevelt. This was basic research in its most Earth-shaking form. The White House tapped Bush with a new, less dramatic, directive: write a report. But this wasn't your ordinary white-paper. It seeded an unprecedented growth of pure inquiry and American prosperity.

In July, 1945, just months after Roosevelt's death, Bush presented President Harry Truman with the written case for government-subsidized basic research. *Science—The Endless Frontier* cites the development of wartime essentials like penicillin, radar, and the atom bomb as evidence that the pursuit of scientific progress had assumed a newly hallowed position in American life. The takeaway was compellingly simple. *Our national security and prosperity depend on scientific progress. Let's keep going.* The report made a splash in the American media and began changing the nation's attitudes toward science—an attitude further enhanced by one of history's ghastlier episodes of uncanny timing. Less than two weeks after the release of *The Endless Frontier,* the Army Air Corps dropped the atomic bomb on Hiroshima. The event shocked the world and cast America as a leading military and scientific authority. Bush's basic science had become nationalized.

The bomb created a quandary. It didn't simply end a war; its detonation also marked a beginning of an era in which wars would be fought with weapons of greatly advanced scientific design. No longer would nations be able to gear up for war in a matter of months. The new weapon proved the value of basic research.

This lesson wasn't lost on America's erstwhile ally, Stalin's Soviet Union, which devoted a vast portion of its economy toward research with a military purpose. The Soviets detonated their first atomic bomb in 1949, while assigning teams of researchers to develop an intercontinental missile.

During this escalation, NACA made headlines by breaking the sound barrier with the Bell X-1, a just-commissioned aircraft capable of achieving supersonic speeds. The committee was still at the mercy of U.S. military objectives, with plans in motion to convert the supersonic airplane into a supersonic bomber. And NACA began working with the man behind the V-2 rocket, Wernher von Braun. The Americans could have had him tried on war crimes; instead, the government invited von Braun and his team to develop a new rocket, one capable of launching a satellite into orbit.

Von Braun had a brilliant counterpart in the Soviet Union, an engineer and physicist named Sergei Korolev. In August, 1957, Korolev and his team sent up an intercontinental ballistic missile and lofted it 4,000 miles from Kazakhstan to Kamchatka. The official news agency put out a bellicose announcement, noting that the launch proved "the possibility of launching missiles into any region of the terrestrial globe." Then, little more than two months later, the Soviets sent Sputnik into orbit. Millions of Americans watched from their backyards as it passed overhead through the night sky. The implication was staggering. Space was, demonstrably, the next frontier. If the Soviets got there first, what advantages might this give them? Senator Lyndon Johnson warned that the Communists would soon "be dropping bombs on us from space like kids dropping rocks onto cars from freeway overpasses." Even before the Americans could send their own satellite into orbit, it appeared that space could become the next battlefront.

But President Dwight Eisenhower saw an alternative. Even before Sputnik, he had authorized a peaceful satellite program. Eisenhower did everything he could to make space a demilitarized zone. But the Cold War got away from him. A few months after Sputnik, a satellite built by the U.S. military, the Explorer-1, was launched into orbit by von Braun's rocket team. Nonetheless, Eisenhower offered a vision to Congress of a peaceful space race; and, in 1958, he proposed creating a new government agency out of NACA. The National Aeronautics and Space Act created NASA. It was the next logical step in Vannevar Bush's push for basic research. Instead of setting up a new government agency for civilian space initiatives, Eisenhower instead chose to subsume and expand a scientific collective that already existed. NACA's former 12-man committee had already expanded to 8,000 scientists and engineers. It was about to get a lot bigger.

The first NASA budget of $100 million reflected the changing priorities driven by the Cold War. In 1961 it began Project Mercury, with the intent of orbiting a human. The agency chose the Atlas rocket, a missile capable of sending a capsule into orbit, manned by a single astronaut. While its purpose was profoundly peaceful, the Atlas had a martial ancestry. The rocket had started out as an Air Force intercontinental-missile program whose roots went back to the German V-2. But within Atlas's history lies a continuing lesson: leading-edge technology requires a long lead time. The space program of the 1960s stood on the shoulders of decades' worth of research and testing.

With the development of Atlas came a program to recruit and train military personnel to become astronauts. But that April, the Soviets pulled another surprise, sending cosmonaut Yuri Gagarin on a single-orbit trip around Earth.

Eisenhower's successor, John Kennedy, had been no fan of space. He saw it as an inefficient use of resources that could be put towards Earthbound projects. In fact, MIT's Jerome Wiesner, the science advisor to both presidents, later suggested that Kennedy would have mothballed the space program if he could have done so without tarnishing his popular reputation. But the news of the Soviet cosmonaut changed his thinking. Kennedy needed, desperately needed, a chance to prove something to the Soviets and the world at large. His attempt to invade Cuba at the Bay of Pigs had been an embarrassing failure. The Soviets were rapidly building their nuclear arsenal. Nikita Khrushchev had threatened to "bury" the United States. So soon after America had led the Allies to victory in a world war, we were beginning to look like the underdog. Kennedy sought some way, short of commencing Armageddon, of looking like the world's leader again.

Initially, Kennedy mulled building massive desalination plants to win the favor of Third World nations. But he wanted something more. Kennedy knew that becoming the main actor on the global stage required more than neighborly acts. Taking care of others' basic human needs—food, water, shelter, medicine—is not sufficient to awe the world. America needed also to demonstrate its *power* in peace, leading the way to an inspiring new high ground. And so he assigned his space-loving vice president, Lyndon Johnson, to look into the prestige-boosting aspects beyond Earth. "Is there a space program we can undertake and win?" Kennedy asked Johnson.

He had come to the right man. Johnson had been the Senate majority leader in 1958 when Congress passed the bill that created NASA. He saw clearly the scientific, economic, and diplomatic potentials of a civilian space program. After meeting with Wernher von Braun, and holding brainstorming sessions with NASA's top leadership, Johnson decided to back a plan that NASA administrator James Webb had proposed not long before: America would send men to the Moon and bring them back safely. It was an ambitious project, requiring a host of technological innovations that had yet to be invented. And the project's cost was high. When Webb had initially brought the plan to the White House, Kennedy had rejected it. This was what Johnson, tasked with finding a program that the United States could "win," took back to the Oval Office. This time Kennedy agreed.

Selling the plan to Congress and the American people required a feat of rhetoric that matched the cause. Remarkably, Kennedy turned the occasion, at the height of the Cold War, into an argument for peace. Instead of beating the Russians with weapons, America would turn an enemy into a rival. We would vault past the Russians and go all the way to the Moon. Kennedy had pivoted with remarkable speed; he pitched his argument to Congress just five months after taking office.

Then, in September 1962, he gave his famous Moon speech at Rice University. The school had helped arrange the donation of land near Houston, Texas, for the purpose of establishing a space center, a generous nod to Kennedy's Texan, space-loving vice president. Addressing a crowd of 35,000 in the Rice football stadium, Kennedy picked up on Eisenhower's vision of a peaceful space, a frontier, as Kennedy put it, that "can be explored and mastered without feeding the fires of war."

Kennedy's rhetoric worked with Congress, which had a Democratic majority and members eager to see space facilities and laboratories built in their own districts. But the argument never entirely won over the American people. A majority of Americans thought we were spending too much money on space. At the height of the attempt to send men to the Moon, when NASA sent three humans 25,000 miles an hour into space, circled the Moon, and brought them back safely, most citizens opposed the program. Even when, on July 20, 1969, Neil Armstrong and Buzz Aldrin walked on the surface of the Moon, almost half of Americans thought the money should have been spent elsewhere. The moral is clear: if not for the Cold War, Apollo would never have gotten off the ground. And if not for Lyndon Johnson, a master of the legislative process, the Sixties may have been spent taking salt out of the developing world's water.

Yet, despite the public's budgetary skepticism, Apollo proved to be a bonanza for technology, and it spun off a host of inventions that permeate our economy to this day. NASA was responsible for developing freeze-dried food for zero-gravity mealtime, cordless tools that could be used to collect Moon samples, digital signal processors to enhance pictures of the Moon (a precursor to the MRI machine), and camera-on-a-chip technology used in smartphones today; along with remote cameras, microphones, and other communications equipment capable of allowing the American people a front row seat to the mission.

I can't resist including a few more Apollo-based innovations. This is just a sampling:

Lightweight mini-computer
GPS
Kidney dialysis machine
Smoke detector
Rechargeable pacemaker
CPAP breathing machine
Advanced prosthetics
Lead paint detector
Improved vehicle brakes

Not only did these inventions quickly find their way into the lives of Americans, many of them produced abundant tax returns for the government itself. The plan catalyzed consumer industries and sent the U.S. economy soaring to unprecedented levels.

The larger, ingenious part of the Apollo payoff was the Moon landing itself. On July 20, 1969, an estimated 600 million people around the globe watched Neil Armstrong take his first steps onto the lunar surface. It was the "where were you?" moment for not one but two generations. The landing firmly re-established the United States as the world leader of basic scientific research, a winning gambit for the late Kennedy and President Johnson. (The Soviets' burning hope for a successful Moon landing—the Soyuz spacecraft—had failed its first flight in 1967, sinking the USSR's lunar efforts.) The mission resounded as a symbolic affirmation of America's superiority as a nation, and it sent a rallying call for further scientific research and innovation. It is no coincidence that many of today's leading advocates for STEM education—including tech luminaries Elon Musk and Jeff Bezos—cite the Apollo landing as a foundational moment. Even before Armstrong and company touched down, the Apollo program had employed thousands of researchers from American universities as part of its R&D effort. By the 1960s, space had become an outlet for practical scientific ventures. It was basic science, manifested as a mainstream career path; the Moon landing would ensure its growth. If Samuel P. Langley had been born again during Eisenhower years, he could have enrolled at Harvard as an astronomy major and found gainful employment upon graduation. If nothing else, it would have fit a man who came of age when there were few careers for stargazers.

Altogether, NASA achieved six successful lunar landings, concluding on December 19, 1972 with the Apollo 17 mission. When the USS *Ticonderoga* fetched the Apollo 17 crew from their command module in the middle of the Pacific Ocean, President Richard Nixon was ready to declare victory and pull the plug on

Apollo. NASA and the last few Apollo crews had raised the bar by deploying a lunar rover and exploring the Moon for multiple days. Nixon's announcement came as a disappointment to the agency. Had the president given NASA more time, astronauts could have been sent to explore the lunar mountains and some of the Moon's most impressive craters.

Space enthusiasts believed that Apollo's giant leap for mankind was just a first step. *Star Trek*, which began broadcasting at the same time Apollo was getting off the ground, seemed like a prophecy. Surely we'd be seeing weekly flights to the Moon by the early Seventies, as von Braun had envisioned. Remember, it had taken a mere 11 years from the Wright brothers' first flight to the first commercial plane trip. The time span between the first transcontinental airmail service to the establishment of the first transcontinental airline? Three years. The span between the Soviets' first rocket and the Americans' Moon landing? Twelve years. With that kind of momentum, what would keep us from continuing straight on to the next planet?

Politics.

President Kennedy had set Apollo in motion to beat the Russians. As long as we were competing with the Soviet Union, the space program could ensure generous funding. But when the Cold War ended, the political fuel for space exploration began to diminish. The historian William Burrows argues that Apollo had been undertaken "for exactly the wrong reason."

Perhaps. On the other hand, maybe it was exactly the right reason for the time. Apollo provided a momentous, peaceful demonstration of American systems at work, in order to persuade other nations to choose our systems. And maybe that still is a good reason for an ambitious space program. Aren't we competing today to influence the peoples of the Middle East, Asia, and Africa? Aren't tyrants and religious extremists using our tarnished reputation against us?

Besides, one would expect that a quest of Apollo's magnitude, attempted by a democracy, would involve politics. The question isn't whether the motives for Apollo were sufficiently noble. The question is whether motives other than war, hot or cold, will drive space policy in the future. And whether we will understand the historical certainty of human settlement of space and choose to lead the way or watch others take the lead.

Richard Nixon offered an answer. His task group on space, led by Vice President Spiro Agnew, understood that the heroism of astronauts would drive support for the program. But space had developed an industry that had become economically indispensable. It had taken on a momentum of its own. The question was how to keep that momentum going in the cheapest way possible. Nixon's decision: a reusable spacecraft that would send people into orbit at the lowest cost. And so was born the Space Shuttle. *Columbia* took off in 1981, completing 28 flights over the next 22 years. It and the four other Shuttles notched a total of 135 missions, launching satellites and carrying out experiments in space, until the program ended in 2011.

Yet, while the bulk of NASA's budget was being spent on human space flight (even while NASA's budget declined dramatically in real dollars), scientists were most excited about the robots being sent into space. Right at the tail end of Apollo, in 1972, an orbiter called Mariner 9 mapped nearly all of Mars and sent back the first photos of the Martian moons, along with a massive amount of data—and tantalizing hints of life. During our Bicentennial, the Viking landers, equipped with a biology instrument to detect life, descended on Mars. They showed good evidence of water and generated data that took two decades to examine.

Meanwhile, other probes were leapfrogging Mars and heading to the outer reaches of the Solar System. The Pioneer series set speed records after they launched and sling-shotted past the Red Planet; Pioneer 10 reached the speed of 82,000 miles an hour. In 1973, it brushed by Jupiter, taking pictures just 81,000 miles above its surface. The probe showed that Jupiter's Great Red Spot is probably a gargantuan storm. Pioneer 10 then soared on the solar wind, passed Neptune's orbit, and in 1983 left the Solar System, where it continued to send back data. Pioneer 11 took more pictures of Jupiter, explored Saturn's rings, and examined the planet's moons. The Mariner series took close looks at Mars. The probes told an alarming story: Venus is far hotter than its position from the Sun would warrant. Consisting almost entirely of carbon dioxide, Venus's atmosphere offered planetary proof of the Greenhouse Effect. The Mariners showed one key practical benefit of planetary exploration.

While the spectacular images from the Pioneers and Mariners—and those of their successor, the Voyagers—roused ecstatic attention from space lovers and scientists, space flight got the most news. In 1986, with the explosion of the *Challenger*, the news couldn't have been worse. Besides the human cost, the Shuttle disaster put NASA on hold for months; and it delayed the launch of one of the agency's greatest triumphs, the Hubble Telescope, by three years.

Paradoxically, the disaster actually increased public support for space. Before the accident, a majority wanted to keep funding for NASA constant. Afterward, 63% supported building a space station, and half said they wanted to send people to Mars. An additional irony: the entire cost of sending probes to explore the other planets, including strange Uranus and our neighbors' many moons, cost about the same as a single Shuttle mission. But the response to *Challenger* shows the essential human reason why manned space flight continued. Without people in space, there would be less funding for the robots. And as NASA associate administrator Frank Martin put it, "We don't throw ticker tape parades for robots."

That feeling got support from a blue-ribbon commission appointed by President Ronald Reagan. Its report's title reflects the chief theme: "Pioneering the Space Frontier." It called for spending a "steady" amount of the federal budget to study space, mine the Moon and Mars's Phobos, and establish a colony on Mars.

The Mars ambition got seconded by Reagan's successor, George H.W. Bush, who called a colony on the planet "inevitable." But the end of the Cold War delayed that ambition. With the absence of a worthy (and scary) rival, Congress cut the funding for a manned mission to Mars.

Still, the hope for a Mars colony continued, with successive presidents up through Barack Obama calling for sending humans there. At the same time, NASA's directives continued. In 1993, the International Space Station got approved, with assembly in low Earth orbit beginning in the late 1990s. The Russians, along with 14 other nations, participated in the construction. Meanwhile, NASA's Mars Rovers crawled across the planet. But the agency's scope increasingly found a mismatch with its funding. The agency told Congress in 2011 that it may not be able to reach space at all in 2016, without better rockets or funding.

President Obama redirected NASA's efforts in space. Henceforth, he said, the government would skip the Moon and head straight to Mars, getting there sometime in the middle of the 2030s. The work in low Earth orbit would be taken over by private industry. Many Americans received the mistaken impression that space was being taken over entirely by private companies, the most notable led by high-tech billionaires. Why mistaken? Because without the government, not one of those companies' rockets ever would have left the ground.

We have come a long way from Langley's day. But the need for the federal government to bear the risk has not ended.

4

Going Private

On December 22, 2015, the first stage of a rocket called Falcon 9 helped drive a payload of 11 satellites into low Earth orbit, then, 35 seconds later, executed a ballet of engine burns and stuck its landing just five miles from the launch pad on Cape Canaveral. The first rocket with an orbital payload that managed to return for a successful upright landing, Falcon 9 was a triumph for its company, SpaceX; for its brilliant Internet billionaire/car maker/space entrepreneur Elon Musk; and for the cause of commercial space. Musk's billionaire/drone maker/ space entrepreneur rival, Amazon founder Jeff Bezos, couldn't resist a me-too Tweet: "Congrats @SpaceX on landing Falcon's suborbital booster stage. Welcome to the club!" Bezos's own space company, Blue Origin, had previously managed to land a booster, but it hadn't carried a payload to orbit.

Still, the moment was significant for more than the technology. SpaceX had succeeded in proving the practicability of relatively low-cost space flight. Blue Origin, taking a more systematic approach, was providing strong competition. And a host of startups were hiring engineers and testing concepts for sending satellites, laboratories, and tourists into low Earth orbit. It was a whole new space race—a race that, if you believed the media, was putting the government's space efforts to shame. A typical headline, from *Inc.* magazine, read: "These Space Entrepreneurs Are Leaving NASA in the Dust."

The news spin left out a critical detail: If it hadn't been for direct government support—and the knowledge ore from NASA's previous missions—the SpaceX, Blue Origin, and Virgin Galactic rockets would never have left the ground.

Elon Musk is a prime example. The man behind the successful launch of PayPal (and the inspiration for Robert Downey Jr.'s portrayal of Iron Man) learned that space is even harder than the

Internet. An achiever accustomed to impossible goals, he was all of 28 years old when he sold his first software company for $309 million. He was 31 when he sold PayPal to eBay for $1.5 billion. Several months before that sale, in June 2002, he had founded SpaceX. The company rapidly built the Falcon 1, named after Star Wars' *Millennium Falcon*; along with the Dragon spacecraft, designed to carry both cargo and humans. Both rocket and capsule were built in just seven years. Impressive, though slower than Musk had hoped. He had planned to put a vehicle on Mars by 2010.

While we all rightly celebrated SpaceX's innovative entrepreneurship, the company's primary role was that of government contractor. Musk sank a total of $100 million of his own money into the company. In 2006, NASA awarded it a $278 million contract to develop technology for servicing the International Space Station; an additional agreement raised the total to $396 million. But the contract stipulated that the rockets and spacecraft actually had to work, and the Falcon's first three launches failed. Meanwhile, SpaceX was spending $4 million a month on unproven technology. By 2008, the company was not only in trouble; it was threatening to bring down Musk's other ventures: the Tesla electric car and SolarCity, a company that makes solar power systems along with charging stations for electric cars. Once again, NASA came through with a boost, awarding SpaceX a $1.6 billion contract to ferry supplies to the International Space Station. NASA orders to date have totaled $3 billion—30 times Musk's personal investment in SpaceX.

That "dust" SpaceX left NASA in? For Elon Musk, it constituted pure pay dirt. The company now seems financially stable, sending rockets up at an average of one a month, carrying payloads for both private industry and government with its Falcon 9 rocket—often from the Vandenberg Air Force Base.

Yet this fact in no way diminishes the scope of SpaceX's achievements. NASA's help shows how government can catalyze entrepreneurship and technical innovation. The agency's ideal roles lie in three areas.

Technical and financial risk: NASA has achieved the biggest firsts, including landing on the Moon and sending probes beyond our Solar System. Private industry, on the other hand, excels at doing well-understood jobs efficiently and quickly, particularly when it comes to repeated tasks like sending satellites into low Earth orbit. One example, the business model of SpaceX, uses ten identical rocket engines in each Falcon 9, requiring just a small team to assemble, reducing cost. Musk wants to send payloads into space for one-tenth the expense of NASA vehicles.

Pure science: NASA is best at the missions that explore the edges of technology and share the results publicly. These missions do not promise an immediate financial payoff. They enable pure science and create a knowledge ore for private companies to mine.

Contracts: As America's biggest customer, the federal government can pump funds into competitive startups, providing an instant guaranteed market. (Remember the Postal contracts in Chapter Three.) Contracts saved SpaceX and ensured its continued survival.

These elements help explain government's role in developing technology. Relatively little tax money has gone toward actually building newly developed equipment like airplanes. Most of the time, the government doesn't build. It buys. By creating a market for the new aviation industry, Congress took the economy airborne. It funded an experiment to prove whether planes were practical for more than warfare. As a result, aviation would boost the American economy and help define the American century.

No wonder NASA saw contracts as an important part of its catalyst role. If the twentieth century was about private aviation, the twenty-first would be about private space. In both cases, government would be on the extreme ends: the leading edge of science and technology, and the back end of contracts and catalysis. Elon Musk himself argues that NASA's budget should be at least double what it is today, as "life insurance" against human extinction. Meanwhile, Musk is still planning a SpaceX Mars landing, in hopes it will inspire more national support for space. In other words, he envisions a public-private partnership.

THE BILLIONAIRE RACE

Musk has good reason to believe in the space program as a source of inspiration. He himself was sucked into space by Apollo. It's one thing that all of the space entrepreneurs of this century—individualists and iconoclasts all—have in common: as children, they were enthralled by America's accomplishments in space.

Burt Rutan, one of the first private space pioneers, was one of them. An aeronautical engineer by training, he started out as a civilian employed by the Air Force before founding his own aircraft company in 1974, specializing in homemade airplane kits. In 1982, he founded Scaled Composites to create aircraft designs. After designing a record-breaking plane that flew around the world without refueling, Rutan and his company went after a loftier goal: to win the Ansari X Prize by sending the first private rocket to carry two people 100 kilometers above Earth twice within two weeks. Rutan accomplished this feat on October 4, 2004, with the backing of Microsoft co-founder Paul Allen. The billionaire joined up with Rutan in a new company called Mojave Aerospace Ventures. Rutan designed a rocket that would be carried by a large airplane (modestly called the White Knight) before igniting its own fuel, thus saving tons of weight. Instead of a parachute, he designed SpaceShipOne in the shape of a badminton shuttlecock, with hinged tails in the rear to slow its descent. The whole project was accomplished by a team of fewer than 50 people and cost a mere $28 million. Rutan bragged that the company developed "an entire manned-space program from scratch—our own rocket motor, our own rocket-test facility, and our own flight simulator for training pilots… with absolutely no help from Nay-Say—excuse me, NASA."

He had a right to sound cocky. Rutan had a reputation for cheapskate innovations, especially when it came to testing technology. Instead of a wind tunnel for his round-the-world plane, he had tested a model on top of his Dodge station wagon. He proved that cost-cutting could work, in contrast to the enormously expensive Space Shuttle program, with its multiple government contracts pushed by members of Congress in individual districts. The Shuttle had been partially a jobs program. Rutan had something else to prove. After he won the Ansari X Prize, he had his picture taken with a spectator holding a sign that read "SpaceShipOne, Government Zero."

There was just one thing wrong with that attitude. Like Newton, Rutan was standing on the shoulders of giants—including the NASA rockets that preceded SpaceShipOne.

Admittedly, the catalyst for SpaceShipOne's dual flight didn't come from government. The Ansari X Prize was backed by an insurance company, with the $2 million premium paid by Anousheh Ansari, an Iranian immigrant and telecom entrepreneur, along with her brother-in-law. But the idea for the prize came from space entrepreneur Peter Diamandis, founder of a company that makes microsatellites. His inspiration for the prize came in turn from the $25,000 Orteig Prize offered for the first successful flight between New York and Paris. When Charles Lindbergh won the Orteig in 1927, he became a global celebrity. There was talk about running him for president. His achievement inspired thousands of young men and women to go into aviation. One aerospace historian wrote recently, "The prize spawned the $250 billion aviation industry."

And yet, as you've seen, the biggest aviation catalyst, the first big initial boost for an entire industry, came in the form of a turgidly written government contract. As for Peter Diamandis, the man who conceived the X Prize: a champion of private space, he relied on a $100 million Defense Department contract to help launch his company.

BECAUSE IT'S COOL

Private space has a lot going for it; not least the unflagging enthusiasm of its technology-born champions, who are themselves inspiring young people. While they vary in age—Burt Rutan is 28 years older than Elon Musk—they share remarkably similar stories of being inspired by space. Nearly all grew up watching

Star Trek. All of them talk about Apollo as a life-changing experience. All of them are STEM geeks, inspired by space to acquire a technical education. (Peter Diamandis being a bit of an outlier, having gone to medical school to please his parents.) And all of them wish to inspire a younger generation to go into space.

In fact, the space industry bears remarkable resemblance to the industry that enriched most space entrepreneurs: the Internet, a space that's out of this world—in this case a virtual world instead of the actual one. Think about it; a generation ago, few people imagined that the "bricks and mortar" world of retail would be rocked by an online bookstore named after a river, or that people would trust their money to flow through a single website called PayPal, or that computer geeks would become billionaire celebrities while still in their twenties.

Those Web and software entrepreneurs defied the establishment in every way possible; Steve Jobs, Bill Gates, and Elon Musk were proud college dropouts who spent their careers defying expectations and achieving the impossible. Musk didn't just become rich in his twenties; he founded and sold two successful companies while in his twenties, and went on to create three companies in three years, all in different industries— space, automotive, and solar energy. This was a group of men and women who seemed capable of achieving anything. Just as Apollo had inspired a generation of technically minded people, the Internet generation attracted the best and the brightest to explore their own virtual worlds. But those Web pioneers themselves were inspired by the space pioneers who went to the Moon. When the time came to put their vast new wealth to the highest purpose possible, entrepreneurs like Bezos, Musk, Ansari, and Allen understandably chose space. It made sense. "Astronauts were like heroes to them," said John Spencer, president of the Space Tourism Society, in an interview with *Entrepreneur* magazine. "Once they grew up and became wealthy enough, they migrated essentially to the space world, because that's the ultimate challenge."

Space is the coolest; the most noble, next-level thing there is, a cornucopia of impossibilities to overcome, just like the early Internet. Like the virtual space of the Web, outer space offers otherworldly exploration. And besides all that, there's money to be made. Vast amounts of money.

There's another tie to the analogy between Web and space: their initial dependence on government. As every modern historian knows, the Internet began as ARPAnet, a computer network among research universities set up by the Defense Department. The original satellites that vastly increased the speed of communications between those computers were sent up by NASA. It's no exaggeration to say that, without the government's original efforts in networking, microelectronics and satellites, Bezos, Musk, Ansari, Allen, and Jobs would have had to make their money elsewhere. There would be no Internet as we know it—at least not yet.

Similarly, as we have seen, the space entrepreneurs have depended on the government's pioneering efforts in R&D, risk-taking, and contracts.

But the point here is not to disparage private space ventures. On the contrary: their initial dependence on NASA and the Defense Department shows just how government can stimulate the next generation's economic and technological innovations. The key point is that we need a change in paradigm. The anti-government pendulum must swing back toward the middle before the private and public sectors can advance more efficiently in space. The American economy, a

phenomenon unprecedented in history, was built by both public and private efforts. Private space ventures are carving out
niches in space, only one or two of which overlap with NASA's
continuing efforts. While Elon Musk continues to set the lofty
goal of a private Mars colony, most of the entrepreneurial
ventures in space are limited to the area between Earth and
lower orbital paths around it. Private ventures focus on five
basic areas; tourism, satellites, research, and—speculatively, at
least—space mining and suborbital travel. Currently, the space
industry alone comprises a $300 billion-plus industry that has
been growing at a healthy annual pace of 4.9%.

TOURISM: HIATUS IN THE HEAVENS

After winning the X Prize, Burt Rutan joined up with Sir
Richard Branson, the flamboyant founder-CEO of the Virgin
Empire. Branson asked Rutan to remodel SpaceShipOne to
carry tourists into space. The new venture, Virgin Galactic,
begun months after Rutan won the X Prize, has already signed
on 700 tourists—including such celebrities as Leonardo
DiCaprio and Lady Gaga—each paying $250,000 a ticket. For
that price, SpaceShipTwo will take them on a two-hour flight
up to 100 kilometers, with the queasy passengers given a total
of six minutes of weightlessness. Virgin Galactic had hoped for
its first paid flight by 2007; but, like most private space programs, the timing has proven more difficult than the optimistic
entrepreneurs anticipated. A SpaceShipTwo pilot was killed in
2014 when a safety design error allowed the co-pilot to pre-
maturely unlock a braking mechanism. The company is now
looking to test a new system in 2017, with no date set for tourists buckling in.

The first tourists have already gone up, just like the man who took that original flight to St. Petersburg. Space Adventures was founded in 1998 by aerospace engineer and website entrepreneur Eric Anderson, with help from his mentor, X Prize founder Peter Diamandis. Space Adventures partnered with the Russian space agency to send tourists up in Soyuz capsules to the International Space Station, for a price tag of up to $30 million. Since 2001, seven people have taken the tour, including Anousheh Ansari. Actually, it's unfair to call all of the passengers tourists. Ansari, for one, conducted four scientific experiments for the European Space Agency during her eight days on the space station.

Another competitor, XCOR, is designing a rocketplane to take tourists on a half-hour flight to 100 kilometers and back again, for the low-low price of $100,000 to $150,000. The company has some serious engineering chops, having designed the first successful rocket-powered plane, the EZ-Rocket, back in 2001. Still, the company has suffered the delays endemic to space. It had hoped to have its Lynx rocketplane fully operational by 2010. While its website has an alluring Book Your Flight button, the company recently laid off half its workforce, with signs that the Lynx will suffer further delays.

Jeff Bezos's Blue Origin website has no booking button. Like XCOR, Blue Origin plans to take tourists on suborbital flights. But the company has not announced a ticket price, instead taking a more systematic, less splashy approach.

On the other hand, hotel magnate Robert Bigelow, founder of Las Vegas-based Budget Suites of America, is focusing on building rooms in space. He started Bigelow Aerospace with the purpose of building inflatable modules for rent by pharmaceutical companies, universities, and others—including, potentially, tourists. He hopes eventually to park a huge space colony, with a capacity for thousands of people, on Lagrangian point L5, using materials mined from the Moon. For starters, the company sent two inflatables into orbit in 2006 and 2007, inspiring NASA to order one as an expansion wing for the International Space Station. The Bigelow Expandable Activity Module, or BEAM, went up in a SpaceX Dragon module to the International Space Station in 2016. The BEAM boasts an interior space of 514 square feet, about the size of a deluxe hotel room. While Bigelow envisions tourists eventually occupying his stations, the estimated price is clearly intended for corporations: $25 million for an extended stay of two months, transportation not included.

The LEO economy is within sight. Private investors are now demonstrating launchers, re-entry spacecraft for humans, and orbital outposts, hotels, and laboratories.

SATELLITES: ORBITAL RUSH HOUR

It really wasn't that long ago when the two greatest superpowers were vying to put satellites into space. Now, 50 nations have their own satellites in low Earth orbit. If you're a Thailand, say, you can call Space Systems/Loral, a Canadian-owned company based in Palo Alto, California, and tell them you want to put a satellite into geostationary orbit for television broadcasting or military communications. You can have the thing in orbit 25,000 miles above Earth within two years. What's more, competitors are pushing that lead time down to eighteen months or less. And the cost is plummeting, all thanks to competition among private industry.

Already the slots in geostationary orbit and low Earth orbit are beginning to get crowded. At last count, 1,469 satellites were in orbit, a third of them American. Currently, competition exists largely because of governments, thanks to the large price of putting a satellite into orbit—at least $50 million and as much as half a billion. Most of that price comes from the cost of the rocket, vehicle, and fuel. Most of the rocket, in fact, is fuel—including the fuel required to carry the fuel. To launch a satellite into low Earth orbit, the rocket must push against gravity and atmosphere to achieve a speed of 17,000 miles an hour. For geostationary orbit, the required speed is 6,875 miles an hour. The greater the weight of the launch vehicle, spacecraft and satellite, the higher the cost. Not long ago, the price per pound of sending an object into low Earth orbit was about $5,000. Now, the Russian Proton-M rocket may have cut that cost by half.

Musk's SpaceX intends to drive the price down still further—far enough to create a true private market in space. One big cost-saving measure is the reusable vertical takeoff and landing (VTOL) rocket. Both SpaceX and Blue Origin have built working VTOL rockets. SpaceX, with its big Falcon Heavy rocket, promises a price per pound in low Earth orbit of $709. Whether or not Musk can achieve that goal, the goal itself is crucial. Private industry is exceptionally good at producing technology that can conduct repeated tasks at a low cost. Whole economies are based on this sort of feat.

Capitalism itself, of course, is based on competition, and the number of companies vying to put satellites into space is growing. Jeff Bezos is one of Musk's most celebrated competitors. But Blue Origin is just one of at least 16 companies, along with three commercial wings of national space agencies, making rockets to launch satellites commercially. Two are European partnerships, several American, three Russian, along with Japanese, Iranian,

Indian, and assorted multi-national firms. All are vying to bring launch costs down.

Another way to lower the cost of satellites is to make them smaller—much smaller. The miniaturization of electronic circuits has allowed for satellites to shrink dramatically. Sputnik, two feet in diameter, could do nothing more than emit radio pulses. The largest satellite currently in space, a top-secret spy satellite launched by the U.S. military in 2010, is floating in geosynchronous orbit over the equator. It required a massive lift by a Delta IV Heavy rocket, capable of 1.9 million pounds of thrust. While the Pentagon will not release the figures, the cost was certainly large.

By contrast, a new market is springing up for nanosatellites, complex equipment weighing just two to 22 pounds. These satellites can be launched individually or in clusters, saving huge amounts of fuel and allowing the development of specialized small rockets to carry them. Then there are the emerging pico-satellites, super lightweights of a couple pounds or less, which can be launched in swarms. Among the most famous is the CubeSat, originally conceived by engineers at California Polytechnic State University and Stanford in 1999. This cube of circuitry and communications equipment, weighing two pounds, can be built and launched for as little as $150,000. A CubeSat can piggyback with larger equipment on a rocket launch, sent out from the International Space Station, or be deployed in groups to communicate with a mother satellite. CubeSats are ideal for high-risk laboratory experiments—science with a relatively low chance of success. NASA recently announced a Cube Quest Challenge to design picosatellites to orbit the Moon in 2018. Five teams, a mixture of universities and private companies, have won $20,000 prizes for completing the first round.

SCIENCE: FLOATING LABORATORIES

The miniaturization of satellites opens up possibilities for less than lavishly funded science departments around the world. But universities, electronics companies, and pharmaceutical corporations already occupy a sizeable payload portion of satellites in space. And not just for exploration of space and Earth. The absence of gravity causes crystals to grow unusually large and with almost perfect form, without touching the walls of their containers. These crystals are prized for everything from smartphone components to drugs.

In addition, space is ideal for X-ray crystallography, which examines the structure of molecules. In just one example, researchers probed the protein that causes Huntington's disease, a terminal degenerative ailment that has yet to yield a cure. The protein is nearly impossible to crystallize on Earth; so in 2014, Caltech researchers sent it up in a SpaceX Dragon capsule to the International Space Station. After the crystals formed in the station's microgravity environment, they were returned to Earth for lab study. The launch was funded by a NASA contract.

Science on the Space Station gets coordinated by CASIS, the Center for the Advancement of Science in Space. The entity has funded and facilitated experiments ranging from vascular tissue research to cell growth on scaffolds (known fondly as "organs on chips"), to fluid dynamics, to remote sensing. CASIS also lets private businesses conduct experiments, and it forms partnerships with nonprofit organizations to spark interest in STEM. For example, Boy Scouts in the Chicago area are readying microgravity experiments for the Space Station, undoubtedly sparking widespread jealousy among budding scientists across the nation.

MINING: PLATINUM ASTEROIDS

Someday, perhaps 30 years from now, companies will be grabbing asteroids and exploiting them for mineral extraction. Asteroids contain the same valuable minerals, from gold to platinum, from cobalt to indium, that space showered upon Earth billions of years ago. One sizeable asteroid can hold $20 trillion worth of minerals. On a smaller, more practical scale, Peter Diamandis estimates that a single 100-foot asteroid can contain as much as $50 billion of platinum. The good news is, we can exploit asteroids without damaging Earth's environment, and the supply is literally endless. The bad news: asteroid mining will be enormously risky and expensive, at least in the beginning.

To gain the promised prosperity, the exploration must be funded by society as a whole: through our government. The leading program, OSIRIS-Rex, a science-driven mission, was due for launch in 2016. The plan is to travel 300,000 miles to asteroid 101955 Bennu and bring back two kilograms (a little more than 4 pounds) for study; the round trip will take seven years. In short, we'll get our first chunk of asteroid in 2023, at a cost of $800 million, rocket not included. (The rocket costs an additional $184 million.) Spurred by President Obama, a more ambitious project has been proposed, to lasso an asteroid and bring it into orbit around the Moon for further study. Its cost of about $2 billion seems out of range to many members of Congress.

Still, as we shall see in the next chapter, that's where government *should* come in. Long-term, high-risk ventures with a literally incalculable payoff are the work of a great nation. As a whole.

EXCESS OPTIMISM

Government's role in space requires patience. Missions take years, even decades. The distance our probes cover is vast. Results are far from instantaneous, the costs are huge, and the payoffs are neither immediate nor always recognizable. It's easy to compare a behemoth like NASA with the much-slimmer, more nimble SpaceX and its competitors. After all, they're run by brilliant aerospace engineers and dotcom entrepreneurs, while NASA is run, ultimately, by 535 members of Congress, many of them more interested in producing jobs in their own districts than in the efficiencies of missions.

But space is humbling for everyone, and it offers harsh lessons to everyone attempting to enter the field. Newly formed space companies tend to offer wildly optimistic predictions. You can see them in an otherwise authoritative book written by a law professor at George Washington University, Lewis D. Solomon. His 2008 book, *The Privatization of Space Exploration*, is filled with promises made by private companies at the time. In 2009, Solomon predicted, Richard Branson's Virgin Galactic would be carrying tourists on suborbital trips. By 2011, Space Adventures intended to offer customers a trip around the Moon on a Russian Soyuz spacecraft. By 2012, Robert Bigelow promised a month-long stay in one of his space hotels. Unlike NASA, he said in a mission statement in 2005, "I'm used to doing things pretty darn well on budget and pretty darn well on time."

These are truly great Americans, and their optimism, even their hubris, is a strength. It takes a rare confidence to boldly go into space. But hubris can become bad policy when the public believes that the private sector alone can accomplish everything in space. Entrepreneurs have an essential, indispensable role. But so does government, which represents the combined resources of *all* citizens.

This inspires me to talk about farming in the next chapter.

STIMULANT

5

The Martian Farm Act

An audacious public space program can bring incalculable benefits—along with some benefits that are not only calculable but fully proven. NASA's work with aeronautics and space has helped improve air transportation, jump-started global communications, boosted technologies in medicine and materials, and provided a jobs multiplier far beyond other sectors.

We rarely hear "audacious" and "public" in the same phrase, but I use them deliberately. In this chapter you'll see how the government can serve as more than a catalyst for the private space industry. It also lends patience to long-term investments in space—a necessity for meeting the long timeframes of missions to other planets and beyond. With the backing of the American people, NASA can spread the risk of early investment in space across society, reducing the uncertainty that keeps wise investors from backing private ventures. Government can go where reasonable venture capitalists fear to tread.

In short, the public American aerospace program reflects our courage in pursuing the future.

"Courage" is yet another word rarely heard in connection with government, outside of the military. But America's most far-reaching economic successes often start with courageous government investment. The twentieth century offers a retrospective map, along with powerful evidence for my argument. The first American century, a period of sustained growth, benefited from periodic structural changes in the economy—changes driven by new industries and new technology. These innovations were catalyzed by government research and development funding, sustained over time. More than that, Americans' investments made through public government agencies have *fostered* whole new markets, industries, and areas of technology; such as nanotechnology, global positioning systems, and, not least, the Internet. These benevolently disruptive technologies did not arise by accident. They originated with Vannevar Bush's vision, and came out of distinct *missions*, scientific, military, and economic, defined by agencies such as NASA. "Transformational public investments were often fruits of 'mission-oriented' policies, aimed at thinking big," such as going to the Moon, writes Mariana Mazzucato in her recent book, *The Entrepreneurial State.*

Courage and patience. Risk-taking. Boldness. Thinking big. All of this language swims against the current political narrative on both the right and the left.

The political right sustains the belief in a direct relationship between small government and economic growth; the lower the taxes, the less active the government, the bigger the economy. In this view, Adam Smith's "Invisible Hand" serves as an unerring self-correcting mechanism, balancing supply and demand, eliminating the lazy and inefficient, and rewarding the risk takers and innovators.

On the left, John Maynard Keynes remains the patron saint of economic theory. The economy is full of flaws, uncertainties and brutal cycles—thanks in part to the irrational behavior of private investors, whose "animal spirits" make them put money into ventures based on the zeitgeist, trends, or personal beliefs. "Market failures" result—including recessions, inequality, and inefficient distribution of resources. Keynesian theory holds that government must serve as a corrector and a counterbalance to animal spirits and market failures. The public sector can stimulate the private sector where needed and can step in where the economy simply doesn't work, such as health care or policing.

Neither side alone shows why America's technological know-how dominated the world's economy during the last century. Can we credit low taxes? In the mid-century, at the peak of the good ol' days often recalled by the more nostalgic members of the right, the top marginal tax rate exceeded 90%. And, as Mazzucato points out, before the 1980s the capital gains tax was more than double what it is today. In other words, the tax burden, especially for wealthy citizens, was far heavier during America's greatest economic boom.

Was it the New Deal and its policies of redistributing wealth? While most economists agree that Franklin Roosevelt's administration helped restart the economy, his more redistributionist policies essentially rearranged existing wealth. Ultimately, though, growth must come from new wealth; and most new wealth comes from technological innovation. Sustained new wealth grows out of new technology, particularly in agriculture and aerospace. These two sectors have provided the largest long-term positive trade balances over the past century. The growth in American agriculture and aerospace could not have happened without bold, strategic investments, through well-defined missions, by the American government.

FARMER ASTRONAUTS

Matt Damon's role in *The Martian* perfectly combines two of the greatest government investments in the economy. His character, Mark Watney, is a botanist-astronaut who must "science the shit" out of his predicament. Left alone on Mars, he sustains himself by growing potatoes on the barren soil. Watney can thank the nineteenth-century Morrill Acts for the know-how behind the sciencing. Farmers in general, and all Americans, can thank the Hatch Act of 1887 for spreading that know-how across the land. (Earth, that is. Mars comes later.)

The Morrill Acts established the land-grant university system by giving federal land to states, which then sold the properties to endow colleges. The law required these colleges to teach a mix of liberal and practical arts, including agriculture, science, military science, and engineering. Thanks to the Morrill Acts, well-endowed public institutions were able to train and hire leading agricultural researchers. The Hatch Act ensured that this science spread to the nation's farmers. It established agricultural experiment stations affiliated with the land-grant colleges; a later law, passed early in the twentieth century, created cooperative extension services in every county to bring innovations directly to farmers. The combination of intelligent law and strategic investment helped turn America into the world's food powerhouse, with a trade surplus since 1960—$39 billion in 2014 alone.

Our success in agriculture offers a critical economic lesson. In "sciencing" farms, the government sparked unprecedented growth in an essential sector. The Agriculture Department did not merely fix a market failure, as Keynesians would have it. Nor did it step aside and wait for American farmers to manage entirely on their own. Government *enabled* farmers to innovate on their own, and to share that knowledge widely. The resulting science-based agricultural network would eventually help feed the world.

It should come as no surprise that the government used a comparable model to help create the aerospace economy. Although the Wright brothers managed to fly the first airplane in 1903, it was the Europeans who rapidly seized on their invention. Planes superior to that of the Wrights flew in France and Denmark in 1906. Three years later, a Frenchman flew across the English Channel. More alarmingly, in 1911 Italy became the first nation to launch military planes for reconnaissance, during its war with Libya.

In fact, it was war that stimulated Congress to take action. Early in World War I, in 1915, Congress funded NASA's predecessor, the National Advisory Committee for Aeronautics, or NACA. (See Chapter Three.) Just as Congress had funded pioneering agricultural research in the previous century, now it turned to institutionalizing discovery and invention in human flight. In 1917, the committee built its first laboratory, Langley, in Hampton, Virginia. Three other labs followed, along with two small test facilities. NACA went on to invent an air duct that revolutionized aircraft engines and that continues to be used in modern automobiles. The NACA cowling, a fairing or cover that reduces engine drag, is a part of some aircraft nine decades after its invention. Similarly, NACA airfoils—essentially, formulas that determine the shape of wings—set the standard for airplane design used to this day. NACA engineers provided critical solutions to the problems Boeing had with the B-17 bomber before World War II, and it provided the British government with a wing design that perfected the Mustang fighter plane built by North American Aviation. Many of NACA's inventions were "bootleg" projects on the part of individuals and small teams; the committee's management encouraged these small side missions with the (correct) expectation that they would lead to greater discoveries.

In general, when Boeing, Douglas or other commercial companies had a problem, they brought it to NACA's Langley Laboratory to solve. Even during my NASA career at Langley, the lab was brought in to solve "911" calls from the likes of Airbus, Northrop, Orbital, and Boeing. NASA engineers worked to solve buffeting problems that McDonnell Douglas was having with the twin tails of the F-18 fighter jet. More calls came in. The composite tail of an Airbus plane had broken off during takeoff. The Pegasus air-launched rocket needed urgent engineering help with aerodynamics and controls. And Boeing's Delta II rocket was suffering instabilities on launch. For these urgent requests for help, Langley engineers serve as the equivalent of the Agriculture Extension Agents who answer farmers' questions about what's devouring their crops.

In short, NACA and its successor have helped push innovation, while also enabling the inventions of commercial industry to succeed.

RUNNING INTERFERENCE

As important as the inventions themselves is their use in private industry. NACA research created a rich vein of knowledge ore, a public intellectual resource that could be tapped by the private sector. In effect, the federal agency ran R&D interference for the nascent aviation industry. To show how integral NACA was to the creation of new aircraft, we need look no further than one of the agency's biggest mistakes. In 1941, as America was frantically seeking to match the Germans' superior technology, Lockheed encountered a potentially devastating problem with its new P-38 Lightning plane. Manufactured at the outset of the war, the plane's tail had a tendency to shake violently during high-speed dives of 522 miles per hour. At the same time, the nose would duck, steepening the dive, while the controls locked up. Lockheed engineers attempted to fix the problem, but a pilot was killed during one test. The lead designer, Kelly Johnson, decided that wind-tunnel experiments would be the only practical path to a solution. But the large wind tunnel at Langley operated at a maximum speed of 100 miles an hour; a newer, much smaller tunnel at the Ames Aeronautical Laboratory in California maxed out at 250 mph, less than half the speed Lockheed required. Johnson begged NACA to ramp up the speeds, but the agency's leadership objected, arguing that the experiments would damage the tunnels. Eventually, the head of Army Air Forces, General Henry Harley "Hap" Arnold, ordered NACA to boost the tunnel speeds. By the time the agency complied, Lockheed had been set back a year of R&D. The experiments worked, however; they showed that the solution lay in dive flaps under the tail. While the story can be seen as one of typical government caution and intransigence, the fact remains that NACA developed the tunnels in the first place, long before the aircraft industry had the technology—or could take the financial risk—to build them. When NACA refused to change the tunnels, Lockheed had nowhere else to go, demonstrating Mazzucato's point better than any economic theory.

NACA went on to lead development and testing of the first supersonic aircraft. A former NACA engineer working for Bell Aircraft first approached the Army in 1944 with the concept of exceeding Mach One. NACA engineer John Stack led the research, using the vast amount of data that had been developed in the NACA wind tunnels. In 1947 the Bell X-1 plane—originally called the NACA XS-1—became the first to exceed the speed of sound in level flight. Two men won the prestigious Collier Trophy in aviation for the achievement: pilot Chuck Yeager, and John Stack. The lesson here seems perfectly clear: Bell Aircraft, in close collaboration with America's aeronautics agency, mined the public knowledge ore created in NACA's wind tunnels.

In 1957, with the Russians' launch of Sputnik, Americans had every right to a sense of déjà vu. Just as the Europeans seemed to take the lead in heavier-than-air flight at the outset of World War I, now the Cold War saw our enemy beat us in orbital flight. Once again, Congress worked under pressure, folding the 43-year-old NACA into a new National Aeronautics and Space Administration. Under presidents Eisenhower and Kennedy, NASA took on a distinctly peaceful bent, despite the Russian threat. While the agency continued to perform military tasks, the bulk of its budget went mostly to peaceful missions, including going to the Moon. NASA not only won the Cold War Moon race, it helped create what's now a $200 billion per year satellite industry, plus the weather-prediction industry ($11 billion per year in direct economic benefit), along with technological innovations that nurtured the microelectronics, communications, and medical-device industries. While it is impossible to make a precise calculation of the exact economic benefits of NACA's and NASA's efforts, no expert would question the general economic benefits and the critical role the agencies have played. Together, these industries comprise 10% of the American economy—while improving the quality of life for billions of people around the world.

NACA and NASA have not followed the government agricultural model, exactly. While the land-grant universities and extension services have operated on a decentralized basis, NASA operates under a central authority, focusing on missions as well as research. But their stories combine to create an important conclusion, one that most politicians and policymakers seem to have forgotten: bold, patient, sustained, risk-spreading investments in innovative technology led directly to two of America's largest continuing economic successes. The public knowledge ore has created the best-fed generations in the history of humanity, avoiding the mass starvation that experts had predicted as the inevitable result of increasing global populations while making it possible for people and goods to leap across continents. Along the way, private investors and entrepreneurs have tapped into that ore and created companies with millions of jobs. Without USDA and NASA, we would not have reached that state. Nor would we have achieved it without private sweat and investment. The right mixture of both holds the secret to our powerful American economy.

The problem is, we increasingly celebrate private efforts while forgetting the public ones—in spite of the evidence and experience. We continue to forget to our peril.

THE RISK LANDSCAPE

Mazzucato, the economist, likes to describe the investment world as a landscape; a "bumpy and complex one," as she puts it. Government's job—when government is visionary and well-run and not dogmatically following Keynesian or libertarian ideology—is to get out ahead and smooth the way. NASA and other agencies invest in new mission-oriented technology "until fear-inducing uncertainty" gradually levels into "mere risk." Economists argue about the difference between uncertainty and risk; but as an engineer and physicist, I like to use numbers to distinguish between the two. In short, risk has numbers. Uncertainty lacks them.

Take the weather as an example. You turn on the radio or check your smartphone before you go to work, and the forecast says there's a 30% chance of rain. That means there is a 70% chance of a nice day. If you leave without an umbrella or raincoat, you risk getting wet. The risk, according to the forecast, is about one in three. You know the odds, they seem to be in your favor, and you take the risk. It has a number. But imagine it's pitch dark outside, you've lost your phone, and your wifi and radio are on the fritz. You can't get the weather forecast and have no idea what it's like outside. Worse, you don't know what the weather will be this evening when you return home from work. There could be a blizzard for all you know. Or a hurricane. The clouds looked pretty weird the night before. But you can't put a number on last night's clouds. That's uncertainty. And what do most people do in the face of uncertainty? They behave more cautiously than they usually would. You probably would be more inclined to bring a raincoat. Or, if you feel a little freaked out about the uncertainty, you might even stay home. Better safe than sorry.

Similarly, imagine you're a venture capitalist investing in an industry where an average three out of every four startups fail. Among the successful companies, just one out of eight doubles its value every year for the first five years. What are your odds of making a good investment if you backed only one company? How about backing four companies? Ten? While the math is complicated, a good computer algorithm can parse the odds precisely. Will it eliminate your risk altogether? No. But it can tell you exactly how much risk you would be taking.

Now suppose you are offered an investment in a market that does not even exist yet—extraterrestrial mining, say. Two entrepreneurs come to your venture capital firm and make a pitch. They want to start a company that captures a small asteroid, parks it at one of the Lagrangian points, and then extracts valuable minerals. Some will get sold to other companies for use in space construction, and some will be brought back to

Earth. "We're talking minerals worth as much as billions!" says one of the entrepreneurs, showing a beautiful slide with that beautiful number on it.

"How much of an investment do you need?" you ask.

The entrepreneurs look a little embarrassed. "Three billion," one says.

"Three billion!"

"Actually, three billion a year," says the second entrepreneur.

"For ten years," says the first.

All right. In this scenario, suppose you actually have the money. Your firm has made a killing in Silicon Valley. But what are the odds that the venture will succeed? Thirty percent? Sixty? You have no idea. There are no odds, because there is no record of companies' success or failure. You have no investment version of a weather forecast. You're completely in the dark. While the idea of vast profit has a certain appeal, your firm would not see that profit for at least ten years. What venture capital company in its right mind would make that kind of investment?

While I made up the eager pair of entrepreneurs, I did not make up the extraterrestrial-mining part. A great many highly intelligent, well-informed people are working up plans for the industry right now. The only problem is, the industry won't exist for at least another two decades. Even five years requires a vast amount of patience for a venture capitalist. And VCs tend to invest at a relatively safe moment in the life cycle of a company, after a market exists and industry is defined. Many startups face a dangerous time, when they have burned through most of their capital and have not yet begun to receive significant income. Economists and investors use a term for this phase, borrowed from the tech professions: the Valley of Death. Most of the big money waits for companies to get through the valley, proving their business model. If markets and industries make up the terrain that includes valleys of death, then government programs allow new terrain to be discovered. Venture capitalists like to help fast-growing young companies with the means to expand, tap a larger market, and then reap the profits when those companies get sold or go public.

New industries have not even reached the Valley of Death. When the brilliant physicist Richard Feynman first conceived of the field of nanotechnology in the 1950s, for example, the term "nanotechnology" had not even been invented yet. (A scientist at a public university in Japan coined the term in 1974, and an American engineer at MIT popularized the concept with a book in 1986.) The nanotech industry is just getting off the ground, with nanomaterials a billionth of a meter thick being employed in sunscreens and carbon-fiber bicycle components, among other uses. But the industry remains in its infancy, and the far-out, science-fiction-worthy products—nanorobots repairing patients' organs, clothing that can change color and shape—require patient, and uncertain, investment over decades. That is why the U.S. government created the National Nanotechnology Initiative in 2001, with an investment totaling $21 billion through 2015 and continuing today. An effort combining universities, private companies, and government agencies (including NASA), the initiative works to create a trove of knowledge ore and pass it to the private sector as a whole, all while helping train a workforce skilled in the new industry.

Extraterrestrial mining and nanotech are two examples that speak to Mazzucato's assertion that the State "often takes the lead, not only to fix markets but to create them." She notes that in biotechnology and the Internet, venture capitalists began investing 15 to 20 years after public funds had been committed to the industries.

THE STORY CHANGES

Yet, even while government and the private sector were working together to build America's greatest economic successes, the narrative increasingly turned against any public involvement in the economy. We celebrate the achievements of Elon Musk and Jeff Bezos while forgetting the public ore they are mining. "We no longer think governments should have missions," Mazzucato told the *Financial Times*.

One reason for this belief is the history that government has of picking the wrong "winners." Opponents of public venture capital point to the Solyndra bankruptcy. Washington guaranteed loans to this manufacturer of thin solar cells made of innovative materials, hoping the company could compete against the common silicon-based photovoltaic cells. However, prices of silicon dropped just as Solyndra was getting established. Private investors pulled out, and the company declared bankruptcy in 2011. Opponents of President Obama used the failure against him during the 2012 presidential campaign, arguing that Solyndra was proof that government shouldn't attempt to pick winners.

Mazzucato has a different interpretation. Investors, whether public or private, always have failures—and public investment failures are necessarily more common than private ones. While private venture capitalists look for commercial viability within a three-year or five-year span—a strategy that tends to kick in after the company gasps through the Valley of Death—public investment properly deals in new fields, where companies are just getting established and where technology is in its earliest, most iterative stage.

What's more, public investment covers more than private startups. The Apollo mission directed 85% of its budget—more than $100 billion—to private companies, creating a robust space market. The risks were not just financial, of course; three astronauts lost their lives to Apollo. Yet the Moon landing was made possible by the earlier, public risks taken in low Earth orbit. The investments and losses of money and life, the successes and failures, smoothed the way for the private space industry we celebrate today.

The arguments against public venture capital come from the left as well as the right. Liberals as well as libertarians argue that tax dollars should not go to make life easier for wealthy capitalists. Mazzucato's answer is interesting. She counters that government, not just private companies and shareholders, should benefit from the successes of public investments. Essentially, she argues that the knowledge ore should not be free. Just as the extraction industry must pay the government for mineral rights on public land, technology industries should pay for extraction of the knowledge ore. The problem we have today, she says, is that "risks are socialized and rewards are privatized." In other words, we taxpayers bear the risks, and private companies reap the rewards. "SpaceX," Mazzacato told a *New York Times* reporter, "is free-riding on NASA technology." If the State were given a share in those rewards, that money could go toward more investments, while covering the inevitable losses of early investment in a nascent industry—much as a diverse portfolio covers its own losses with its own gains. The technology development funded by government should pay for itself.

How exactly it should pay has raised skepticism among Mazzucato's critics. She hints at giving agencies an equity share in startups such as Tesla or Google that depend on government-created technology. The critics doubt the practicability of determining the value of that dependency. How much is the technology worth at various stages of the company's growth?

On the other hand, suppose government patented its inventions. The critics respond that industries would grow more slowly as a result. They offer proof in the Human Genome Project, which made all of its gene sequences public at no cost. The project's main private competitor, Celera Genomics, patented its discoveries. As a result, companies developed products from Celera's genes at a much slower rate than from the public genes. Another danger: Once the government becomes accountable for making money from its own investments, what is to stop it from measuring success in the same way as a private investor? Would a government primed for profits become as risk-averse and impatient as any sensible venture capitalist?

More important than who pays for the big government risk is the lack of awareness that government is taking risks in the first place. Private industry used to swing for the technological fences. Giant corporations—most notably Bell Labs (the research arm of AT&T), Xerox PARC (innovators of the interface technologies behind the computer mouse), IBM's Thomas Watson Lab, Alcoa Research Lab, and DuPont made a series of impressive discoveries throughout the first half of the twentieth century. These companies could transform their discoveries into products and then take them to market; this business model no longer exists. These days, the private sector funds just 18% of basic research in this country. The public sector funds

57%, with universities and foundations covering the rest. And Congress is cutting public funds by billions per year.

We're not just talking about science but technology as well. Gregory Tassey, an economist with the National Institutes of Standards and Technology, warns of "long-term inadequacy" in research that enhances productivity in technology, physical, human, and organizational capital. Tassey points out that, while building housing creates jobs and produces a multiplier effect—spreading wealth throughout the community and beyond—technology creates much more leverage and spreads wealth throughout a variety of industries. Besides, he adds, technology jobs pay much better. Technological advances contribute unquestionable value, creating whole industries and advancing whole economies—particularly ours. Perhaps the tax money is worth it. As Mazzucato herself points out, when Eisenhower created NASA, the highest earners in America were paying a top marginal tax rate of more than 90%; it is now just over 39%. Capital gains taxes have been cut by more than half since the 1980s. It is a hard fact to swallow, but the economic successes that have come from NASA were paid for with taxes—our own as well as those paid by the industries themselves. The question is whether the money is worth it, not just in advancing humanity but in actual economic benefits.

Space itself offers an immediate answer. It provides a fundamentally new economic field. Just as most citizens and leaders in 1900 could not possibly envision the size and importance of today's air transportation market, most of us have trouble imagining the scope of the future space transportation market. Yet that market is already being built to provide lower-cost launch services, tourism, manufacturing, and rapid intercontinental transportation. Private investors are beginning to inch into the market created by government investment of our tax dollars. Will this new industry succeed? There will be failures, just as there was during NACA's long, patient investment in aeronautics. But there will be brilliant successes as well—provided NASA's most active constituencies start working together to change the narrative. The future cannot rely exclusively on narrow efforts aimed at near horizons of three to five years.

Now let's look closer at that technology that stimulates the economy.

6

Deep Tech

Non-scientists tend to think of science (when they think of it at all) as a series of brilliant thoughts punctuated by bolt-out-of-the-blue discoveries. Think Newton under the gravity-prone apple tree. Or Archimedes plunging into his over-full bathtub. Never mind that someone had to invent that tub in the first place, along with the means of filling it with water for his body to displace. Even modern science, we think, gets done by, well, scientists. Who created the Hadron Collider, that huge structure in Switzerland that smashes atoms for science? The obvious (and mostly wrong) answer is scientists.

The truth is, much of the complex equipment used in science requires an equally complex partnership between scientists and engineers—professionals who understand materials, stress forces, energy flows, the effects of heat, cold, light, acceleration, shaking... everything that makes equipment go right, or fail.

When it comes to space, engineers' role seems even more profound. Not only do they lead the invention of the tools of science itself, they invent the tools that take the science tools into space. And those tools lead directly to useful, profitable inventions on the ground, often in unrelated industries.

In other words, the discoveries of science depend on invention. The creations of engineers lead to discovery, which in turn helps lead to more creation. As an engineer and a physicist, I have spent my career pursuing inventions and discoveries. Actually, the ancient Greeks and Romans saw the two as much the same thing. In the Old Testament, the writer who calls himself Ecclesiastes—the one who said "there is nothing new under the Sun"—was stating a common belief of the ancients. All knowledge already exists, fully known by the gods, they said. It only awaits discovery by us mortals. Marvelous new devices— new kinds of ships, new weapons, improved metals—came from intelligent humans bringing together the means available to them. Imagine a clever Greek inventing an improved mining shovel. It lets workers dig farther and deeper until they hit an undiscovered ore. The ore only seems new to them; it has been lying there for millions of years. Another Greek experiments with this metal, combining it with others, and invents a new, even better shovel. And so on.

Call it the *invention-discovery cycle*. A new tool leads to new knowledge, which leads to a new tool. Over time, the cycle speeds up, first with a burst of creativity in Asia; then with the Enlightenment and the Industrial Age in Europe; leading up to the twentieth century and the invention of the airplane, transistor, and silicon chip. All of these inventions depended on the work of scientists. In turn, the scientists depended on increasingly sophisticated laboratory and field equipment. Throughout the 1900s, as the cycle spun faster and faster, the demand grew for ever more sophisticated equipment. While individuals or small groups could still invent and discover, the cycle came to depend on larger teams and bigger funding. Not even the titans of industry could have gotten us to the Moon. That effort took the backing of a nation.

To this day, the space program constitutes one of the greatest accelerants of the invention-discovery cycle. Besides advancing scientific knowledge, space's contributions to the cycle offer more immediate, tangible benefits. Think of the cycle as a literal wheel of science and engineering: the faster it spins, the more it flings off information that other inventors can use. In this chapter, we'll explore how that cycle works with the space program. Far from being an accidental system, it's a highly controlled, meticulously planned set of ventures.

CRYSTALS IN SPACE

Many branches of science—including Earth science, planetary science, astrophysics, and microgravity science—depend strongly on NASA and its ability to enable experiments in space, or the observation of Earth from space. Every ten years, the U.S. Congress funds the National Academy of Sciences to conduct a survey probing what the scientists as a whole community believe are the next big discoveries. Where's the edge of their science? What are the questions that need to be answered to push back that edge? The report, called the Decadal Survey, comes back to NASA in the form of recommendations. These in turn provide a road map that determines how the agency will spend its science funds. Some money will go to researchers to utilize existing, but not fully explored, data. Some funds will go toward building instruments to gather more data. Other money will fund rockets to launch those instruments. Each branch of science gets assigned a NASA program manager who decides where the money goes. Periodically, a call will go out to the scientific community inviting proposals for experiments. Researchers who have been thinking about how to better understand Earth's radiation budget, or crystal growth, or how stars evolve over their lifetimes will propose experiments or measurements that test their hypotheses.

Imagine a Dr. Mary, a young assistant professor of chemistry at the University of Maryland. Mary wants to combine a set of different molecules to form a crystal, one that she believes might offer unparalleled insulating qualities. Her problem is, her Earth-bound laboratory can't produce the kind of perfect crystal that she needs to prove her hypothesis.

Gravity has a tendency to distort the way crystals form. Dr. Mary sees an Announcement of Opportunity (AO) from the NASA Microgravity Program Manager. She decides to propose an experiment in the International Space Station to grow her crystal in zero gravity, with an astronaut taking pictures of the results.

But it's not enough simply to describe her experiment. She needs to come up with the means to conduct it in space, with equipment that an astronaut can use. What's more, that equipment has to meet standards beyond normal laboratory devices; for one thing, it has to be able to withstand the forces of a rocket launch. It has to fit into the tiny space allotted in a crowded capsule. And it must work perfectly the first time; no one is going to fix it, and there are no spare parts to fix it with.

In other words, Dr. Mary needs an engineer. She happens to be in a good place. The University of Maryland has an excellent engineering department. NASA's own Goddard Space Flight Center sits right next door. Essentially, the engineer (or engineers) must agree to invent the equipment on spec. At this point they're at the proposal stage, and other teams are competing to get their own experiments into space. Engineers at Raytheon, one of the nation's top aerospace contractors—also nearby in Maryland—agree to take on the speculative job. With their help, Dr. Mary's proposal gets the Program Director's approval, and she becomes the proud principal investigator in her crystal experiment. The equipment takes months to build, and an astronaut practices the experiment on the ground for another period of months (along with hundreds of other experiments; astronauts' schedules in space are packed). Finally, perhaps three years after she first responded to the Announcement of Opportunity, Dr. Mary watches online as a Soyuz rocket bearing a capsule with the astronaut and her experiment launch from the Baikonur Cosmodrome in Kazakhstan. A few weeks later, she gets back the pictures from her experiment: perfect crystals!

Eventually, Dr. Mary gets more funding through the National Nanotechnology Initiative, and she works with engineers and materials scientists to develop a new form of heat shield out of her crystal for industry and aerospace. The shield will someday end up protecting a capsule bearing astronauts and scientists as they hurtle through the Martian atmosphere. And the cycle continues: invention leading to discovery leading to invention.

While it's appropriate to credit the government for Dr. Mary's part in it, it's easy to miss the role of private industry. A significant part of the funding in her chemistry department at the University of Maryland comes from the private sector and foundations. A phalanx of companies helped build the International Space Station—Boeing, Lockheed, Raytheon, Northrop Grumman, Orbital ATK, and others. Still more companies are creating the inventions that will take us to Mars in the 2030s, if we decide to fund NASA properly. In fact, most of the work done for space, while funded by government, isn't actually performed by government. NASA's budget provides for about 65,000 jobs, of which only 18,000 are NASA employees.

In short, the invention-discovery cycle is also a public-private cycle.

THE SPINOFF CYCLE

As a scientist, our fictional Dr. Mary was motivated by curiosity. She wanted to test her hypothesis about a novel combination of molecules. But her discovery helped lead to a new invention.

In fact, it's not implausible that the equipment her engineering collaborators at Raytheon invented would find use in future space experiments. The practical knowledge from her discovery and the Raytheon inventions fall under the rubric of "spinoffs." Although the term fits nicely into the concept of the invention-discovery cycle, *spinoff* is modest, even misleading. Spinoffs aren't just the bonus of the space program; they're a critical part of the program itself. We go into space in part for the spinoffs.

Skeptics dismiss this justification for the space program, calling it the "Tang argument." *Hey, without space, we wouldn't have Tang!* Never mind that this statement actually isn't true; Tang had been on the market for three years before John Glenn used it on *Mercury*. At any rate, the use of a powdered fruit drink as an exemplar of the space program misses the point of the invention-discovery cycle. Space aims at the boundaries of knowledge, not just minor consumer improvements. That's why *spinoff* is misleading. The inventions that arise from the space program, even those that find use in unrelated sectors, are no accidents. They're integral to the government's mission.

NASA itself tends to undercut its own argument with its annual report, *Spinoff*. One tempting picture from the report's online version shows glasses of beer with the caption, "A system for creating rocket fuel on Mars is putting the bubbles into beer." While this sounds like Tang all over again, the actual story is far more powerful. Robert Zubrin, the legendary founder of Pioneer Astronautics and Pioneer Energy, has been leading the invention of technologies that convert the Martian atmosphere into water, oxygen, and fuel. Because the air on Mars comprises 96% carbon dioxide, Zubrin has pioneered ways to convert gases into useful substances. For example, he originally speculated that hydrogen could be brought from Earth and combined with Martian CO_2 to make methane and water. Besides providing liquid to mix with astronauts' Tang, they could electrolyze the water to produce oxygen for breathing. The excess oxygen, combined with methane, can make rocket fuel to power the astronaut's flight back home. Now that water has been discovered on Mars, hydrogen can be produced on the planet's surface, and combined with CO_2 to make methane. While inventing the technology that helps make Martian missions possible, Zubin's companies have spun off inventions that produce gas for research balloons. His work with gases may also lead to technology that efficiently extracts hydrogen from water on Earth, producing carbon-free energy. And, yes, Pioneer Energy has created a system that recovers CO_2 during beer making, allowing microbrewers an inexpensive way to put the bubbles back in their beer. It is indeed a spinoff, one that may not advance or save humankind on the order of interplanetary colonies, but... beer!

Put it this way: If Tang could create rocket fuel on Mars and help alleviate climate change, perhaps the skeptics wouldn't scoff as much. Robert Zubrin's beer bubbles are that sort of Tang.

The most powerful spinoffs contribute to the invention-discovery cycle, even while boosting the economy. A perfect example is the work contracted by NASA's Glenn Research Center with MicroLink Devices, Inc. Future space exploration missions may depend on massive solar arrays to provide limitless energy. The problem is, while the cost of solar cells has been declining, they remain expensive. Illinois-based MicroLink has developed flexible cells that can be mass-produced at a fraction of the cost of existing technology. Besides offering promise for space missions, the military has been using the cells for power supplies and in powering drones. Could the technology exist without the space program? Yes, probably. But NASA provides an accelerant and a proven mission-driven invention machine. The Glenn contract changes the uncertainty of solar technology into a reasonable business risk for MicroLink. In return, NASA speeds up the invention of huge solar wings that can send instruments and humans to Mars and beyond.

The spinoffs that don't directly go back into the space invention-discovery cycle nonetheless play a big role in the economy, and in improving our well-being. Not all of them have to do with inventions you can hold in your hand. NASA was a pioneer in the development of systems science—the development of methods to manage enormously complex operations. (In Chapter Seven, we'll explore the ways that policymakers can think like systems engineers.)

The development of systems science and engineering over the decades went hand in hand with improvements in software engineering. The code written for NASA in turn has formed a rich ore of data that companies have mined in unrelated sectors. For example, NASA's Ames Research Center provided software that allows other kinds of systems to detect problems. CEMSol LLC, a Phoenix-based company that creates healthcare software, licensed system-monitoring software from Ames that uses data to track components of health-care systems, establish a baseline for normal behavior, and watch for any deviations from that behavior. The software was designed to actually warn airline pilots and mechanics of possible impending aircraft failures. CEMSol's Integrated System Health Management programs have been used in medical centers around the world.

Among the "spinoffs" from NASA are its engineers, who go on to found private companies. Alliance Spacesystems is one of them. In creating robotic arms for the Mars rovers *Spirit* and *Opportunity*, the former NASA engineers wanted to make it easier to collaborate with computer-aided-design documents. They came up with a PDF collaboration code—and spun off another company, Bluebeam Software Inc., to market it. Bluebeam recently sold to a German tech company for $100 million. How does the software add to the invention-discovery cycle? Engineers throughout NASA—and everywhere else, for that matter—now have a relatively inexpensive, efficient way to exchange and mark up PDF design documents, making invention that much easier.

Similarly, back in the Eighties, NASA software engineer Craig Collier wrote software at the Langley Research Center to help design a hypersonic spaceplane. While the plane itself never got off the ground, the software did, through the private Collier Research Corporation. Collier perfected his code, called HyperSizer, which allows engineers to model weight and load requirements for various vehicle designs. Recently, the company expanded from commercial aircraft design to optimizing wind turbines.

While many spinoffs continue with NASA engineers founding their own lucrative companies, some of the greatest side benefits of the space program work in the opposite direction. NASA contracts with private companies to invent solutions to aerospace problems; the companies in turn figure out other, profitable ways to use their inventions. The Goddard Space Flight Center had one such problem when it came time to repair the Hubble Space Telescope in 1993. Hubble had launched in 1990 at a cost of $2.5 billion, and astronomers had high hopes for unprecedented images of deep space. But when the telescope deployed, they discovered to their dismay that a construction error had left the mirror slightly misshapen, rendering the images distorted and blurry. NASA needed a precise way to detect defects so that properly measured replacement parts could fix the scope. Goddard sent out a call to optics companies, and Massachusetts-based AOA Xinetics won the bid with a detection tool. Xinetics, now owned by aerospace giant Northrop Grumman, went on to create a commercial 3D imaging detection device bought by FedEx, UPS, and just about every other major package shipper.

Goddard had another challenge in 2008. It needed a way to transfer power to electromagnets in space, where the temperature approaches absolute zero. Metal components tend to waste heat, spilling it into space and lowering a magnet's capacity. Goddard turned to the Tai-Yang Research Company, based in Knoxville, Tennessee, to develop superconductive ceramic leads that eliminate waste energy in space. Having created solutions for Goddard, Tai-Yang went on to sell the technology for MRI machines.

Some of the most interesting innovations in materials science are happening with ceramics, nonmetallic substances that can be composed in a huge variety to create insulators and super-hard objects. In addition, engineers at Boston Applied Technologies Inc. (BATi) created a ceramic laser switch that greatly alters properties of light—a function that has uses in telecommunications, lidar, and remote sensing. In 2002, NASA's Small Business Innovation Research Program—a sort of government spinoff generator—provided funds enabling BATi to prove that its new OptoCeramic could work almost

100 times as powerfully as more traditional options. BATi has since produced a catalog of products based on this technology.

Here we're back to NASA's mission-oriented role as business stimulator, which can work even if the initial mission fails, as Craig Collier's work with the spaceplane shows. We could see the plane as a failure, or rather as part of an iterative process that led to aircraft manufacturers doing most of their design of new aircraft with computer models instead of building expensive prototypes. Even better, we can see that the technology that goes into a particular mission rarely stops with that mission. It spins off, allowing companies to get past the Valley of Death stage and to bring new products to market. This role may be the most misunderstood aspect of NASA and other mission-oriented government agencies. Government missions actually work best in areas of high uncertainty, such as space. Failures not only get spread throughout society, where they can be absorbed without shock to a particular sector or system; those "failures" often turn into business successes.

The seeming wastefulness of NASA got encapsulated in an urban legend about the so-called space pen. As the story goes, the Americans spent millions to develop a pen that would work in zero gravity. Meanwhile, the Russians used a pencil. The truth is much more illustrative of the space program. Both the Americans and the Russians used pencils in the beginning of human space flight. But pencils are problematic in zero gravity; the graphite tips can break off and produce dangerous dust, and the wood in pencils can present a fire hazard. So NASA and the Soviets independently went to work developing a space pen. NASA gave up after the effort proved too costly—less than $1 million but a lot for a pen. Meanwhile, inspired by the NASA attempt, pen manufacturer Paul Fisher invented a pressurized ink cartridge and marketed an Astronaut Space Pen. NASA bought about 400 of them in 1967. The Soviets followed, and to this day both American astronauts and Russian cosmonauts use Fisher Space Pens in space. While the NASA failure didn't directly lead to a spinoff, the agency didn't waste millions. And the space program ended up doing one of the things it does best: inspiring inventors.

TRANSFERRING TECH

NASA does not leave spinoffs to serendipity. Ordered by Congress to serve a stimulating function, the agency has a technology transfer (T2) program designed to push out the information ore to the private sector as rapidly as possible. Its T2 website, technology.nasa.gov, serves as a kind of engineering eBay, a catalogue of tempting technologies available for licensing. Any company specializing in optics, software, communications, medicine, robotics, batteries, power generation, electronics, instrumentation, fluid systems, coatings, sensors, materials, and the environment can find pre-packaged patents to adapt for private development.

Maybe it's the engineer in me, but I imagine that anyone who loves technology would enjoy perusing the patent portfolio. Even the names of the patents are irresistible. How can you not click on "Polarization Dependent Whispering Gallery Modes in Microspheres"? These optical structures allow extremely sensitive response to narrow wavelengths, similar to a whispering gallery, which "ignores" other sounds to convey a conversation. Dielectric microspheres, developed at NASA's Glenn Research Center, offer miniaturized uses for sensors in aircraft, health monitors, and optical communications.

Glenn also developed a NanoWire glass switch that uses one-sixtieth the power of conventional microelectromechanical systems, allowing low-cost, low-energy GPS systems and RFID devices—those gadgets that eliminate toll booths on highways and create smart keys. In fact, NASA's extensive work with antennas (obviously an important aspect of space missions) has accelerated RFID development. The Johnson Space Center has invented a new technology that greatly extends the range of RFID. The system promises to improve location systems for first responders, enable hospitals to track patients more efficiently, and help shippers track packages in bad weather.

Anyone with a strand of geek DNA will want to click on the Robotics section of the patent portfolio. The Walk & Roll Robot combines wheels with pivoting hip and knee joints enabling the robot to traverse all sorts of terrains without expending a lot of energy. Plus it can turn at high speeds. Besides potential uses in urban search and rescue and scientific exploration, it could potentially become one of the coolest mechanical toy pets ever invented.

Not all NASA robots are self-contained (although the Robonaut series—human-sized humanoids—can already work side-by-side with astronauts). The Robo-Glove, developed at the Johnson Space Center in collaboration with General Motors, is a lightweight glove that allows a strong grasp with minimal effort, and offers feedback to the touch. The glove offers promise in a wide range of uses, from construction and manufacturing to medicine.

The patents range from the cool to the potentially transformative. NASA's work on battery technology, a critical part of space exploration, could help solve one of the biggest challenges of renewable energy. Unlike fossil fuels, solar and wind power fail to generate continuous energy. We continue to use energy when the Sun goes down and the wind stops. The obvious, and elusive, answer is to store excess energy for use when needed. While batteries have improved over the past few decades, the advances have not come close to what we have seen in other areas of electronics—computer chips in particular. The more energy a battery can store, the larger and heavier the battery gets. This is a particular challenge with space missions, where rockets must carry equipment beyond the Gravity Well. One solution is to make batteries operate more efficiently, by sensing and regulating energy flows. NASA has developed a bevy of patents that allow the management of a variety of batteries.

Some of the patents do not seem immediately applicable to the consumer. For instance, NASA's Jet Propulsion Laboratory created a Deep-Space Positioning System—a sort of GPS without the global aspect—that allows a vehicle to determine its position relative to another object in the Solar System. The system allows navigation without having to establish an expensive infrastructure. Who will end up licensing this technology? Companies interested in prospecting asteroids, among others. Within a few decades, the line between Earth-based industries and space industries will blur. New discoveries in space will lead to new inventions, which will lead to new discoveries. And the cycle will continue to accelerate. But only if we, as a nation, continue to push into the space frontier.

NASA's catalysis of new markets and industries, combined with the knowledge ore from its mission-driven technology, generate jobs and tax revenues that dwarf the agency's budget.

Another significant payoff: developing the human capital we need to innovate.

7

STEM's Roots

When you think of America, do you picture a waving flag? Amber waves of grain? A beer commercial? As an engineer, I think of America as a system. This does not necessarily make me an insufferable geek. My career, spent creating and building first-of-a-kind, has-to-work-the-first-time and operate-in-extreme-environments space robots, has taught me that systems thinking helps us navigate complex problems.

So let's look at America as a system—one that works remarkably well, despite what you hear during election seasons. Our 50 states include several that would rank among the world's top nations in economic might. We grow enough to feed our 300 million people and ship the rest around the world. Our corporations operate globally. Our university system is the envy of the world. Our highways connect us conveniently, our air traffic system safely allows millions of business and pleasure passengers a day to zip around the continent. The power grid is amazingly well behaved and reliable. Businesses operate with ease across state lines because of a massive communication infrastructure. Hospitals, law enforcement, federal agencies, and the military all contribute to a peaceful, long-lived citizenry. From individuals to families to businesses to corporations to state and local governments, America comprises a single complex system.

So what, exactly, is a system? It's a set of parts that work together in a certain way over time and interact with an environment. A system comprises processes, materials or information, and inflows and outflows. How does that definition help you solve problems? Because proven techniques have been developed, partly in response to difficult aerospace problems, that can be applied to real world entities. Originally developed to tame the difficulties of missiles and rockets, these solutions also apply to many other areas. Imagine this scenario:

You're driving your old classic car with a partner on your way to work, through a hot, barren desert. The commute is ridiculously long; lately it has stretched to more than an hour each way, in part because the ancient car can go only 40 miles an hour. What's worse, it lacks an air conditioner. You both roll down the windows, but that barely helps. By the time you get to work each day, you're dripping with sweat. Your trip back home is even hotter. You could stand it if the breeze were stronger or the commute weren't as long.

"We need air conditioning," your partner says.

Here's a classic illustration of the failure to think systematically. Systems experts call this flaw "bounded rationality." It's what happens when you have limited information, or fail to think beyond the bounds of the immediate problem. When I oversaw projects at NASA, I encouraged the engineers under me to think beyond their specialties. A NASA electrical engineer can design you just about any kind of circuit, but it might be useless if it interfered with the vehicle's mechanical system, or used too much energy, or took up too much room, or weighed too much, or used rare minerals that ate up the budget.

You see the same problem with your miserable commute. Following your partner's advice, you put a compressor under the hood and hook a belt up to the engine. This drives the compressor to cool down the car. Woo! Cool air! But now your car goes even slower, and your round trip stretches to four hours. That compressor pulls power out of the engine. Plus, cooling the car's interior decreases the gas mileage, risking running out of fuel in the middle of the desert. Not a good plan. Instead, let's look at the problem systematically.

First, with any system, you have to make sure that any solution improves the system's *function*. Is the car's chief function to cool you down? Of course not. The car is supposed to get you to work.

Second, look to see what parts or interactions are interfering with the system's function. A systems expert will tell you to look for the best places to intervene. You want to seek the smallest change that would make the biggest difference. This often isn't as easy as it sounds. Often, as the late Dartmouth Professor Donella Meadows pointed out, people will "push the change in the wrong direction." Installing air conditioning, thus slowing the car and risking a breakdown, does exactly the opposite of the car's function of getting you to work.

When you look at your car as a system, you realize that, with your AC, you're looking at the wrong part of the car. Maybe it needs a tune-up, or a new carburetor, or a whole new engine. But a systems expert will tell you that you still might not be thinking broadly enough. Your rationality might still be bounded. Sometimes the best solution is not to change a part but to change the system's function—an act that systems people call a "paradigm shift." What if the car's purpose wasn't to take you across the desert? What if you just asked it to get you ten miles to the grocery store? You and your partner set up a new business out of the home. Which has air conditioning.

THE AMERICAN SYSTEM

While my little tale illustrates the trap of narrow thinking, it can also serve as a valid metaphor for America as a system. As with a car, we need to think of our nation as a complex assemblage of interacting parts, within an environment. America's environment is the world. Within that system are subsystems: education, the economy, international policy, the military. Even the government is just another subsystem, or part, fueled by taxes, fees, and borrowing.

What are the main functions of America and its government? Here's where ideology and politics have their say; but not entirely, from a systems standpoint. A system's function isn't what a car salesman or politician tells you. It's what that system actually does. If our classic car were instead a late-model Cadillac, its function, or purpose, would be more than getting you to work. It would get you to work in luxury.

Look at America's most notable characteristics, and you'll find its critical functions. For example, we have a dominant military, the largest economy (including international corporations and the most innovative sole proprietorships), we generate the most patents, we invented the iPhone and online shopping (and the Internet, for that matter), boast the most advanced research universities, and make the most popular music and movies. Given these traits, you might say America's key functions are to be powerful, rich, innovative, and creative.

So what is the function of our federal government? Our politicians' rhetoric would lead us to believe that the government exists to keep us safe from terrorists, foster job creation, and reduce inequality, depending on which party they represent. These are legitimate political issues, and you could argue that each problem could lower the nation's performance as the world's dominant power and economy. But, since we're looking at the government as a subsystem of a larger system, we need to step back and see whether the politicians are talking about the biggest flaws in the American system, whether those flaws are fixable, whether the fixes are the right ones, and whether they screw up another subsystem.

It's hard out there for a systems analyst. Or a president, for that matter. Thinking narrowly is far easier, because it makes the solutions seem that much simpler. But when our leaders start thinking systematically, I believe we'll all be better off.

To back up my point, we need two more systems terms. They're not jargon; both are familiar terms, though you don't usually see politicians use them. The first term is *resilience*. The second, *injection points*.

A resilient system bounces back from any shock or threat. A ponderosa forest in the Southwest gets devastated by fire, and within a few years you see a sea of green seedlings, from pine cones opened up by the heat of the fire. Resilience at its best. In a resilient economy, you often see the fastest growth after a recession burns away the weakest businesses.

Sometimes, though, a system faces a shock so large that it threatens resilience. A forest fire in a tangle of dead trees and branches accumulated over decades can burn so hot that it destroys the pine cones and seeds. In that case, foresters will intervene, with small controlled burns and logging that removes some of the fuel. (The worst fires out West were caused in part by decades of fire suppression—another case of bounded rationality, or failing to see the forest for the trees.) Some systematic problems fix themselves, while others form feedback loops that spiral out of control. On the other hand, sometimes a relatively small, well-timed, deft intervention at the right *injection point* will set everything working again. A tune-up, a new carburetor: the right injection at the perfect injection point.

Back to the American system and its subsystem, the federal government. What problems might hold back our chief functions? What's clogging the system from being the world's most powerful, rich, innovative, and creative? And what injection point could offer the most effective, efficient fix, at the least cost, with the least harm to the other subsystems?

INJECTION POINT: TECHNOLOGY

I wrote a long beginning to this chapter, with its emphasis on systematic thinking, because I've found it difficult to shift policy conversations away from what politicians, talk show hosts, and Beltway insiders love to obsess about. As important as the discussions may be, they almost always fail to fix our fundamental problems. What really makes America great? And what threatens its decline?

You undoubtedly have your own definition of greatness. But I prefer to focus on what America actually does, better than any other nation—in terms of security, innovation, prosperity, and creativity. Just for a moment, accept these characteristics as key definitions of American greatness, arising out of the way America actually functions in the world. You can argue that we're not as powerful as we used to be, that our economy is faltering, that we don't innovate enough, and that African rap is better. But sheer firepower, dollars, box office, and music downloads objectively show dominance if not "greatness."

Now, given these characteristics and the problems that threaten them, what part of the system could be fixed to make most of these characteristics function better? What fix might make America more resilient?

What's the injection point?

To an engineer like me, the answer is obvious: technology. Our military has the best technology. Our economy's growth has been driven by technological innovation for a century. Technology lies behind our military success, our agriculture and aerospace leadership, the iPhone, Amazon, GPS, music downloads, the music itself, as well as our increasingly computer-generated movies. I'm not saying that technology causes all these things, or that tech is the one driver of America's greatness. But it's the one trait that improves every part and contributes to the system's resilience.

Given that tech is the injection point, what exactly is the fix? For that matter, what's the problem?

At first, there doesn't seem to be a problem. The job market in American tech is growing faster than the non-tech market. From 2004 to 2014—not exactly boom times—the overall job market grew by a sluggish 4.5%, while tech jobs expanded by 31%. Technology outpaced even fast-growing sectors like health care and business services. Include all the jobs in STEM (science, technology, engineering, and math), and the future looks especially bright. The Bureau of Labor Statistics projected growth in STEM jobs of 14% between 2010 and 2020; so far, that projection seems right on track.

STEM jobs pay far better than non-STEM jobs—twice as well, in fact. The median annual salary for STEM jobs in 2014 was $78,610, more than double the $33,900 median salary for non-STEM (the mean salary for STEM was $85,570, and it was $47,230 for non-STEM). Granted, despite the growth, STEM jobs make up only 6.2% of the national workforce, or about 8.6 million people. But these same people come up with the greatest innovations and creations, and they add to the nation's military power and its prestige abroad. They drive the economy forward and help create more jobs. They're the leading edge of the American system.

So what is the problem? It lies with the Americans who make our technology; or, rather, the severe, growing shortage of technologically educated and trained Americans. Despite the large growth in STEM jobs, few American students are receiving the education and inspiration they need to fill these jobs. Only 16% of American high school students are interested in a STEM career and proficient in math. Of the students who pursue a college major in a STEM discipline, only about half decide to work in a STEM career. The United States ranks a dismal 34th among industrialized nations in math, and 27th in science.

The picture gets even worse with women. While females make up about 48% of the total workforce, they comprise just 24% of STEM workers—even while women attend college at a higher rate than men. When women do study STEM, their education often fails to lead to a STEM career. Women with a college degree in a STEM discipline are less likely to work in a STEM occupation than their male counterparts; these women are more likely to work in education and healthcare. The Department of Commerce lists "a lack of female role models, gender stereotyping, and less family-friendly flexibility in the STEM fields" as possible factors contributing to the gender gap. Providing more opportunities for women to work in STEM is widely seen as one of the most effective ways for the United States to remain competitive in the global market. Less than 45% of STEM college degrees are received by woman and minorities, though they comprise 70% of college students. And STEM jobs not requiring a college degree, such as aerospace technicians, also are going unfilled.

All of this adds up to too few Americans preparing for, and taking, STEM jobs. The White House projects that by 2018 there will be 2.4 million unfilled STEM jobs in the United States. While an estimated 1.4 million U.S. science-related jobs will exist by 2020, American college graduates are expected to fill less than a third of them.

To improve STEM education among citizens, in 2013 the Obama administration outlined a five-year strategic plan urging $700 million in public-private partnerships. This is a paltry amount in a country that spends $810 billion a year on education. But even if the government doubled STEM funding, would the money double the number of Americans who study it? Would it double America's power, innovation, and creation? Doubtful.

Granted, in many cases the problem may actually be insufficient funding. STEM educational opportunities differ significantly in various socioeconomic and geographic regions. Only 81% of Asian-American high school students attend a school that offers the full range of math and science courses as defined by the U.S. Department of Education (Algebra 1, Algebra II, geometry, calculus, biology, chemistry, and physics). Only 71% of white American high school students have access to the full range of these courses, and the access to math and science courses is much worse for Hispanic, black, American Indian, and Native Alaskan students. Computer science is taught in just one of four high schools nationwide.

But while additional funding would help push more young Americans into STEM, especially where they lack the opportunity for study, the most efficient kind of injection for the federal government to make isn't money. It's something else. The ideal strategy wouldn't push students into STEM. It would create a *pull*, a demand among individuals and communities. If there's a will, any kid destined to be a scientist or engineer will find a way. Any community that takes STEM seriously will find a way. After all, other nations are providing STEM education at a fraction of what we spend per student. The countries that compete with us the most successfully in science and technology—and the comers who might overtake us in the future—are those nations that place the highest priorities on STEM. Not just on the government level, but culturally. Science and technology aren't just valued in these countries. They think STEM is actually cool.

There was a time when that was true in this country as well. Strangely enough, it started with a competitor, Russia, and its small, beeping satellite. It came as a shock, and went on to prove America's technological resilience.

SPUTNIK, THE STEM MACHINE

When the space race began in 1957 with the launch of Sputnik 1, the American public had already entered a new era of technological marvels: affordable automobiles, color televisions becoming the norm, the interstate highway system, more than 1,000 computers built and sold, the development of a polio vaccine. We were on top of the world, technologically.

And then came Sputnik, beating us in getting beyond the world, with technology that our enemy could someday use to destroy us. Simon Ramo, an American engineer who led the development of microwave and missile technology, wrote in *The Business of Science* (1988), "the American response to the accomplishment of the Soviet Union was comparable to the reaction I could remember to Lindbergh's landing in France, the Japanese bombing of Pearl Harbor, and Franklin D. Roosevelt's death."

The response was as profound as the shock, and not just in increased defense spending on technology. For the first time, the nation saw technological education as a form of defense. The National Defense Education Act, or NDEA, increased funding for education at all levels by more than $1 billion, introduced low-interest student loans for higher education, and added a bevy of scholarship opportunities. Curricula in public schools became more challenging, with a new emphasis on math and science. Educators introduced many teaching tools still in use today, including hands-on laboratory classes, overhead projectors, and educational films. NDEA also provided millions of college scholarships; partly as a result, the number of enrolled students more than doubled, from 3.6 million in 1960 to 7.5 million in 1970.

The space program did more than prove our technological leadership. It drew bright kids like high-IQ moths to the light. Franklin Chang-Díaz remembers looking for Sputnik in Costa Rica when he was seven years old and being inspired to devote his life to space. After being told in a letter from Wernher von Braun, then-director of the Marshall Space Flight Center, that he needed to study engineering and learn to fly if he wanted to be an astronaut, Chang-Díaz emigrated to the United States at 18, received a B.S. in mechanical engineering and a doctorate in physics, became an American citizen, and now holds the record (along with Jerry L. Ross) for most spaceflights—seven shuttle missions.

Apollo went on to have an even more profound effect on a generation; witness the high-tech billionaires flocking into low

Earth orbit today. But the lure of Apollo wasn't all about space. It was about the future.

I believe that, in part because today we lack a robust public space program, Americans' attitude toward the future has changed. The future has always been linked to our identity as a nation. We could always make things better, and whatever it took, we were going to do that work. But now that we have achieved beyond the wildest dreams of our ancestors, the rate of acceleration seems to be slowing. We hear evidence of our decline all the time from our politicians. In systems theory, this is called a "drift to low performance." We get used to a lower bar, and we work less to exceed it.

Of course, education is all about the future, by preparing young people. But, as Apollo proved, the best path to the future is a shining one, powered by inspiration. As the saying goes, the best way to predict the future is to create it.

This book has set out to dispel the myth that the space program costs too much. If you look at space from a systems perspective, it seems startlingly cheap—from the perspective of STEM alone. The reason is inspiration. Set out to put people on Mars in 20 years, and you'll have all the scientists and technologists you need for a fraction of what it would cost you to double our investment in STEM. Why would that happen? Because kids are inspired by space to study science, technology, engineering, and math.

More importantly, our best future doesn't happen without STEM. We create the future by educating those who want to create it. The question is, why should they want to? Coolness precedes prestige. Space makes STEM cool. In other words, we need to suck kids into space. If you need proof, get yourself invited to a school when an astronaut is visiting. No event draws more excitement.

And inspiration is cheaper. Doubling the nation's expenditure on STEM education would cost an additional $400 billion a year. Doubling the funding of the space program (and I'm not proposing nearly that much) would cost $18 billion.

Meanwhile, politicians promise the best economy America can buy. Or fulfilling, high-paying jobs distributed broadly among regions, ethnic groups and genders. Strangely enough, the seemingly impossible task of going to Mars, with all the things in the way? That happens to be the easiest, most achievable path to a future we can look forward to. Every six year old knows it. By focusing on avoiding the future we dread, we fail to create the one we want. We've mislaid the good old future, the one we couldn't wait to see. That future is still possible. Six year olds saw it 20 years ago, and six year olds saw it 20 years before that.

DRIFT TO LOW PERFORMANCE

As the space program's goals got pulled back by politicians trying to "save" money, and a post-Apollo generation came of age, the impetus of STEM declined with it. Among top-performing high school students in 1992, just 29% went on to major in STEM subjects in college—far lower than in other developed nations. By the year 2000, the proportion of American STEM majors had dropped to 14%. By 2011, only 30% of high school seniors could even meet college enrollment standards in science. (More than half met reading standards.) An October 2013 survey of more than 150,000 people aged 16 to 65 conducted by the Organization for Economic Cooperation and Development ranked the U.S. 21st out of 23 countries in math and 17th out of 19 in problem solving.

The lack of interest in STEM subjects was reflected in America's high school teachers. Even those willing to teach STEM were increasingly unqualified to teach the subjects. By 2000, 31% of math students were taught by a teacher without a mathematics degree or certification. Two-thirds of physics teachers lacked a background in physics, and 79% of Earth and space science teachers taught without certification. In 2010, the Council of Advisors on Science and Technology declared a crisis, urging the federal government to launch 200 new STEM-focused high schools and 800 STEM elementary and middle schools by 2020—leaving a massive chicken-and-egg problem: the lack of STEM teachers to staff these schools.

And yet, even while our students turn away from STEM, America continues to lead the world in research universities. Eight out of the top 10 schools on the Academic Ranking of World Universities are in America, including the top four. Of the top 50 universities worldwide, 33 are American. At least we can be assured that our older students can avail themselves of a first-class STEM education. Except that they're not, for the most part. The biggest consumers of American universities' postgraduate STEM education are non-Americans. The trend is driven in part by the rising affluent class in China and large scholarships offered to students in oil-rich Gulf states who can get into American universities. Public universities are also recruiting students from abroad to take advantage of the higher tuition rates international students pay. In 2015, 1.13 million foreign students studied in the United States at every school level, a 14% increase from 2014, a 50% increase from 2010, and an 85% increase from 2005. Of these students, 331,371, or 29%, are Chinese. Nearly 81,000 are Saudi Arabian, up from 5,000 in 2000.

Of these 1.13 million foreign students, 974,926 are attending U.S. colleges and universities. In the 2014-2015 school year, international students in American universities had the highest rate of growth in 35 years. The United States hosts more international college and university students than any other country in the world, almost double the number hosted by second-place Britain.

Most of these foreign students are not here for French poetry. Even though the roughly 975,000 international college students only make up about 4.8% of the total number of college students, more than half of advanced STEM degrees from American universities are earned by international students.

This is worth repeating: one out of 20 students in American colleges is a citizen of another country; more than half of all master's degrees and Ph.D.'s in STEM subjects go to foreign citizens. Non-Americans earn 57% of engineering doctoral degrees, 53% of computer and information sciences doctoral degrees, 50% of mathematics and statistics doctoral degrees, 49% of engineering-tech and engineering-related doctoral degrees, and 40% of doctorates in physical sciences and science technologies.

We should be proud of our ability to educate the world. But what kind of power are we giving to other nations? When we decline to a post-Apollo attitude toward STEM, what's the eventual cost?

I mean "cost" literally, in terms of dollars, as you'll see in the next chapter.

8

High Ground

When you first read the preface of this book, you may have checked the cover. Isn't this supposed to be about space? What on Earth does our 50,000-year-old man have to do with sending robots and astronauts to Mars? What do they have to do with colonizing asteroids and planets? And, most of all, what do they have to do with America's reputation in the world?

Everything.

What we're doing in space, and what we need to do, are exactly what has kept humans evolving since the dawn of our species. Our ability to keep going allowed us to travel farther faster than most other species, while bringing our technology along with us. Along the way we learned to hunt, not just cooperatively but systematically, organizing our behavior in common around increasingly sophisticated plans.

Risk. Sustained effort. Superior, constantly improving technology. The ability to exploit environments, to travel from one environment to another, to bring supplies. The ability to carry out complex operations, cooperatively. The certainty that risk and effort would bring great rewards. These are the characteristics that allowed us to evolve at a speed far beyond what genes and environment would normally predict. Our traits are not just in our DNA. They lie in the very definition of what it means to be human. They constitute the story of humanity, which continues...

Ever since our ancestors crossed the land bridge, America has drawn the restless, the inventive, and the explorer. America was a place where treasures could be discovered, brought back by risk takers, and sold to the wealthy. The most fashionable Europeans created *wunderkammern,* "wonder rooms," filled with exotica from America: stuffed creatures, sketches of wild landscapes, portraits of painted and strangely dressed natives. The exploration of America by Europeans helped stimulate unprecedented growth in the European economies. To fund the risky and expensive voyages, governments invested heavily in ships, crews, and military expeditions. Private investors exploited the opportunity by forming one of the great economic innovations of all time, a precursor of the modern corporation: the joint stock company. The ships returned with newly discovered plants, including tomato, potato, corn, and tobacco, along with a trove of silver worth billions in current value.

Perhaps even more importantly, the New World offered a dramatic choice: to stay home or seek the promise of this unknown land. Europe during the early settlement of America was being torn apart by rival powers and warring religions. France, Germany, and England lost thousands of civilians to terrorism by religious zealots. On top of that, overpopulation in some areas, degradation of farmland in others, and deep economic divisions between rich and poor caused misery, disease, and mass starvation. America was more than a land to explore; it was an escape hatch.

Today, when we talk about America as a land of freedom, we tend to think about freedom (or the lack of it) from taxes and regulation, or we celebrate our Bill of Rights. The early settlers in America—those who came voluntarily—sought freedom of a more literal nature. They were escaping religious terrorism, government intolerance, and, often, the bondage of serfdom or poverty. This independent spirit led to the American spirit, long before the Revolution. Americans were colonizers in much the way we can imagine colonizers on Mars. The early voyages from Europe actually took longer than a trip to Mars will take. Like future Mars colonists, the American colonists and successive immigrants had to bring their own supplies and learn how to sustain themselves in a new environment. Unlike future Martians, the American colonists had scarce communications with the homeland; and they faced the understandable resentment of humans who had settled the lands centuries before.

The American colonists, in other words, provide a precedent for our ventures beyond the Gravity Well. More than that, they created an example for the world, of a place that represented the leading edge of innovation, invention, and discovery. Only recently have we risked losing that reputation. Why?

FREDERICK JACKSON TURNER, SPACEMAN

Throughout the nineteenth century, the most restless Americans continued to expand outward, creating a culture unlike any in Europe, albeit one that devastated the cultures it ran across. In 1893, the historian Frederick Jackson Turner wrote an influential paper declaring that American-style democracy, a radical reinvention of the classically inspired republic envisioned by the Founding Fathers, had been created by the expanding frontier. Every time people moved 50 or 100 miles from the farthest settlement, they left behind old rules and customs, Turner said. The American spirit was renewed every time a trapper or a family picked up and moved west.

At the time Turner wrote his famous thesis, however, the frontier was being settled rapidly. In fact, the 1890 census had officially declared that the frontier no longer existed. Americans had gone all the way west and had established scattered settlements in every blank space left (blank except for Native Americans, that is). Without a frontier, Turner surmised, the American spirit would diminish. The restless and the explorer would no longer determine the evolution of the nation's culture.

Turner could not realize that, ten years after he wrote his thesis, the Wright brothers would enter a new frontier of an entirely different dimension: the air. Aeronautics created a different species of frontier mentality; not Turner's separation from civilization but a frontier that tied the nation, and eventually the world, together. America continued to lead in invention and discovery, of a technological nature. The boundaries of flight kept getting pushed back.

American dominance in the air became literal during the Second World War, when U.S. manufacturers cranked out 300,000 planes in just four years. The acceleration of production was unprecedented, ramping up from just 3,000 planes in 1939. Aeronautics leaped from 41st place among American manufacturers to first place during the war. The rest of the world couldn't help but be impressed. America's air forces helped win the war and make the United States the world's most powerful nation. Still, we shouldn't miss the larger lesson: American energy and restlessness created the planes and the technology.

PEACEFUL LEADERSHIP

Ironically, it took the leader of Nazi Germany's rocket program, Wernher von Braun, to jumpstart America's space effort. The German intercontinental missile, the V-2 (V for *Vergeltungswaffe*, or vengeance weapon), began striking England and Belgium in the fall of 1944. More than 1,400 of these rockets hit London alone. The famous broadcaster Edward R. Murrow admired the German "skill and ingenuity" that went into these revolutionary weapons. Six months later, von Braun and several hundred of his rocketeer colleagues surrendered to the Americans.

The Germanization of American rocketry owed entirely to competition with the Soviets, who had made their own effort to entice von Braun and his colleagues. It was natural that a weapon as powerful as the world's first intercontinental missile should be envied by every great power. But when Americans first interrogated him, von Braun suggested an alternative to weaponry. The V-2, he said, was actually a rocket intended for the stratosphere. Of course, he was speaking as a man who might well have deserved to be tried for war crimes. But he had the advantage of being one of aeronautics' greatest visionaries. A more advanced version of the V-2, he said, would lead us to the Moon and eventually to other planets; and he had the plans for those missions.

But equally visionary—arguably more so—were President Kennedy and Vice President Johnson and their advisors. Their decision to put the full weight of the White House behind a mission to the Moon spelled not only a major push for technology. At the outset of the Cold War, facing an existential threat, the American leaders chose space as a psychological weapon.

In retrospect, we can think of the problem of the Cold War in the form of a Q&A:

Q: How can America prove its capitalist democracy to be superior to Communism?

A: Defeat the Communists in battle.

This was the obvious, but foolish, answer. The advent of the nuclear age made that answer self-destructive. The psychological war finally became clear as the best alternative. So the question changed:

Q: How can America win over the neutral nations' people?

The question hearkens back to the American Civil War. As the saying goes, President Lincoln did not just win the war, he won the argument. Kennedy's task, as he saw it, was to win the argument over which system of government, which ideology, was superior. This meant more than putting forth an attractive-sounding political philosophy, and even more than having Hollywood make better movies. To win the argument with the Soviet Union, America needed a single, obvious proof, instantly understandable by every culture, and indisputable by any rival propaganda machine. Something awe-inspiring, heroic, the product of an irresistible culture.

A: Apollo.

The idea that the best way to beat the Soviets wasn't with the military but by accomplishing a miracle was far from obvious. In fact, it was brilliant. Kennedy's speech before a crowd of 35,000 in the Rice University stadium makes this key aim obvious. America had been a lead pioneer in the industrial revolution and the nuclear age. The next big frontier, space, had the "eyes of the world" upon it, he said; "and we have vowed that we shall not see it governed by a hostile flag of conquest, but by a banner of freedom and peace."

Several months earlier, Kennedy made the case even more overtly to Congress:

> Finally, if we are to win the battle that is now going on around the world between freedom and tyranny, the dramatic achievements in space which occurred in recent weeks should have

made clear to us all, as did the Sputnik in 1957, the impact
of this adventure on the minds of men everywhere, who are
attempting to make a determination of which road they
should take.

The impact of this adventure on minds: The Apollo mission was
one of the greatest thought experiments in history. How well
it worked can be judged by historians and social scientists. But
the Moon landings were celebrated around the world, and they
continue to reverberate. To this day, when I speak with people
in other nations about the American space program, they say
how "we" landed on the Moon. Not "Americans." All we humans.

During his Rice University speech, Kennedy used a word
that stands out over the competitive rhetoric of the Cold War:
"share." While the success of going into space "enhances our stat-
ure," he said, space "is not merely a race." The space program's
meaning must be shared: "We go into space because whatever
mankind must undertake, free men must fully share."

America's willingness to share grew more concrete after
the Iron Curtain fell and the Soviet Union broke up. In 1993,
President Bill Clinton agreed with Russian President Boris
Yeltsin to form a joint commission promoting cooperation in
various areas of business, defense, the environment—and space.
Led by Vice President Al Gore and Russian Prime Minister
Victor Chernomyrdin, the commission drafted an agreement
to build the International Space Station. (The European Union
joined the agreement soon after.)

I traveled to Russia as part of the Gore-Chernomyrdin pact
to understand how NASA could employ Russian experts in
optics and lasers. These Russians were more like the Americans
I grew up with than any other nationality I had met. They

were self-reliant, resourceful, and neither accustomed to nor comfortable with luxuries. But they were hurting. One well-educated scientist told me he was making just $30 a month.

The pact had a pragmatic purpose; by sharing the cost, the Americans reduced their own burden. More importantly, the Space Station kept Russian scientists and engineers engaged in peaceful pursuits. Russian experts were ripe for employment by North Korea, Iran, Libya, and other dangerous regimes eager to tap Soviet-era technology and the people who created it.

While the Gore-Chernomyrdin Commission proved effective in using space as a tool in international affairs, it represented a different level of imagination and systems thinking from what the Kennedy Administration exhibited. Kennedy and Johnson had the idea that the best way to beat the Soviets was not so much with the use of the military, but by proving our superior system—especially to the post-colonial states around the globe. The effort was more difficult than the Space Station. And more visionary.

Today, we have added terrorism to rogue states like North Korea and Syria among our greatest foreign threats. Terrorists are not just an enemy force, with soldiers and logistics and a communications network. They represent an ideology as well. Drone strikes kill individuals. Thought and belief, on the other hand, cannot be bombed. We face the same challenge that Lincoln and Kennedy confronted: to win not just the war but the argument.

Islamicists argue that the western system is corrupt and that it's spreading our corruption to their region. American air strikes and military bases in the region may be necessary, but they have the side effect of bolstering the radicals' argument. American aggression, even in the cause of fighting terrorism, proves that we are, indeed, aggressors. We need to return to the view of space that Kennedy espoused: as an adventure that we lead, with meaning that we share. We must show that our creative force can extend into space, not just into certain Islamic nations. That we're not merely aggressive, and not hopelessly corrupt. Cool technology is an attractant, cool to young people everywhere.

Another important reason to fund, and share, the adventure of human space flight is similar to what we offered Russian technologists in the Nineties: an opportunity for employment. A trained engineer or chemist without hope is ripe for work producing technology in the cause of terrorism. National Public Radio unintentionally proved this point recently, in a story about a Syrian refugee welcomed by Sweden. The story reported that the Swedish government paired him with an old-time logger who worked with horses. The Syrian man had never seen a forest before, and here he was being taught useful employment in a new land. NPR presented this story in a positive light. But the reporter revealed that the refugee had been trained in his homeland as an electrical engineer. I'm an electrical engineer myself. Although I'm sure the Syrian appreciated living in a safe place, I can't imagine the man thinking horse logging to be the best use of his double-E degree.

SPACE AS A TOOL

This is not to say that space will solve the problem of terrorism. In fact, I don't believe the space program should be applied in the same way in every region. We should look at each problematic part of the world separately and think how we can use space as a tool for improving relations and gaining respect.

While I am no expert in international relations, I know good systems thinking when I see it. Look at the success of Apollo. Experts in international affairs need to find ways to use space; but they should use space in part as a way of thinking outside their own toolbox.

In other words, we need to apply systems thinking, using space as a tool for improving our international standing in various parts of the world. A systems theorist will tell you that the function of a system is what it does, not what it's advertised as doing. A mission statement or national anthem, or even a Constitution, may not determine a nation's function. The function is what it actually does. Think: What does America actually do? America provides the most opportunity for the most people, while propping up the world's economy. Those are its functions.

Now let's look systematically at how America influences the world. How can the system work better? These functions are much easier when we are the most admired country in the world—or at least the country that makes other countries the most jealous. Every nation, if asked who was number one, would—if they gave an honest answer—say America. It's the most powerful, dominant country in the world. They don't measure GDP or happiness or even transit systems. They look at power and influence, and we're number one. If that indeed is our function—to exert power and influence—then we need to look at the inflection points in the American system. What is the cheapest, most effective way to increase our power and influence? Our greatest influence comes through inspiration rather than coercion. If we look at our system functionally, systematically, unpoetically, we can make the case for space.

Inspiration is a part of that influence. Every marketer will tell you that. Sugar water is sold around the world for billions of dollars because of that kind of great storytelling. Don't discount the value of inspiration.

Think of space as attractive real estate. Imagine if we invited other countries to participate in mining asteroids, and colonizing Mars, and sharing in discoveries, and sharing in the riches that come from those discoveries. On the other hand, what if we provided young people in troubled countries with an attractive alternative to the dead-end tribal, violent belief systems they get from ISIS?

THE VALUE

But let's get more directly practical and see if we can attach a dollar value to the international benefits of the public space program. The amount spent each year on international affairs by the federal government totals about $800 billion a year. That includes the State Department (about $50 billion), Defense ($600 billion), Energy ($24 million), and Veterans Affairs $150 billion). What if a fraction of that $800 billion went to the space program to help build our international standing?

To help answer that question, we need to look at what improves our standing in the world. Most experts point to two factors: Force and influence. Force means military or financial power. Influence is the ability to build hope and opportunity globally, as well as to inspire awe and make individuals and nations want to join us. Let's assume for the sake of argument that half of America's standing owes to force, and half to influence. Let's further assume that America's civil space program—its evidence of our technological abilities, the opportunities for education and employment, and the cooperative efforts led by America with other nations—accounts for a tenth of the total influence.

Eight hundred billion divided by two yields $400 billion. That's America's "influence" budget. A tenth of that—the space program's share of America's influence—amounts to $40 billion. If you accept my numbers (you might choose to weigh the factors differently), then you get a figure considerably larger than NASA's actual budget of $18 billion. Ignore the benefits to the economy, which we explored in Chapter Five. Ignore space's ability to stimulate the growth of STEM education and inspire students to study science, technology, engineering, and math. The international benefits of NASA alone constitute far more value than its current budget.

Admittedly, it's impossible to calculate a dollar value to our standing among other nations. After all, how do we measure greatness? Instead, I let Congress do the numbers for us; and it has chosen to spend $800 billion annually on our ability to protect ourselves, command respect, and influence the world. Space clearly contributes to that influence. Calculate the numbers as you will, and you will find NASA to be a bargain.

Yet, the less tangible international benefits of space, the incalculable rewards, may be of the greatest long-term worth. America has always represented freedom from tyranny and the ability to pursue opportunities in life. While military power is necessary, it is not sufficient to sustain these principles. The influence we gain through peaceful actions, and through the release of creative energy, represents the most effective argument for our system. And nothing makes that argument better than our space program. It transcends the false borders—political, economic, religious, and ethnic—that divide humanity.

In short, the space program, better than any other government program, lets America be America. Space reminds the world, including us Americans, of who we are. We have been the leading edge of humanity's restlessness, the continuing story of the man who first left his home fire, of the people who seek, invent, and discover.

Space is the leading edge of our leading edge. We can choose to let others take the lead. But space lets us truly be ourselves as a nation; and when we are characteristically American—looking outward, pursuing discovery with energy and courage—everyone else looks on in admiration. Space is the epitome of the America the world can't help but envy, and love.

VISION

9

The Quest

On March 7, 1970, President Richard Nixon gave a speech that changed the direction of the American space program. "We must think of [space activities] as part of a continuing process... and not as a series of separate leaps, each requiring a massive concentration of energy," he said. "Space expenditures must take their proper place within a rigorous system of national priorities." While his words seem to reflect my own systems approach, Nixon was actually demoting space as a national priority. Rather than occupying a privileged place in the federal budget, he said that the space program must "be planned in conjunction with all of the other undertakings which are important to us." In other words, space must compete with other priorities.

Fair enough. But the decision to back away from the larger ambitions of space meant throwing away our ability to send people to the Moon. It meant throwing away all the equipment we need to land and operate on the surface. And it meant mothballing the most powerful rocket ever built, the Saturn V. All for the sake of saving money for other priorities. I believe that future generations will see the decision to turn back from the Moon as an act of sheer lunacy—one akin to the Chinese burning their world-dominating fleet of ships in the early fifteenth century.

Yet it would be a mistake to place the blame for the decision solely on President Nixon. Most Americans supported it. After all, many thought, *we were done*. We had been to the Moon. We achieved the goal for the sake of achieving it, without a larger, compelling context. The next great destination would be Mars; unlike the Moon, the planet could not possibly be reached within a decade. Besides, Neil Armstrong and his fellow astronauts had set foot on solid extraterrestrial ground. The enormous feat of getting people to Mars and back would accomplish essentially the same thing: people setting foot on extraterrestrial ground.

We had already met an astonishing, audacious, mind-bending goal. Meanwhile, we forgot what Apollo had achieved besides the Moon. The program had stimulated the economy, fostered the next generation of technology, and inspired future dotcom billionaires to set up shop in the Gravity Well.

After Apollo came the Space Shuttle, with the goal of creating a cargo truck to low Earth orbit. One of the most complex machines ever built by humans, it was also conceived without an overarching vision regarding the program's national benefits. The Space Shuttle helped build and maintain the International Space Station. The ISS, at least, had a larger, noble purpose: to bring many nations of the world together in a common venture. And yet both programs have come to be seen as a dead end. They have fostered great technology and led to scientific discoveries. But Americans aren't being shown the benefits of these space programs besides those of space itself. We don't see the advantages to America, right here, right now.

Our space goals remain orphans, with singular missions denied a larger destiny, lying outside a positive direction for our nation. Americans need more than an admirable scientific or technological feat. We need an even bigger quest—one that, at the same time, reaps immediate rewards. What could that possibly be?

I have been describing it throughout this book. The answer lies in the Gravity Well. It *is* the Gravity Well.

The Well allows Americans a definable path to our future, one that meets our historic destiny as a people. It offers us a unifying principle around the solutions to our greatest challenges: technological, economic, and political. Once we see the Gravity Well in all its challenging glory, it will serve as one of our greatest sources of national inspiration.

While it's easy to see the Gravity Well as a mere trope—the force holding us back, a container we must escape, a challenge just begging to be met—it is no mere metaphor. It's a *thing:* a physical force we feel every time we climb the stairs. It's a definable place, no less real than the American West. And it's a set of goals and bodies in space so real that crashing into one would kill you.

Let's look at the three essential aspects of the Gravity Well.

1. A PHYSICAL WELL, OR A WELL OF PHYSICS

To an astrophysicist, the Gravity Well is a physical force, a field of attraction between two bodies in space, diminishing with distance. For engineers, the Gravity Well is the great physical challenge of overcoming that force. A vehicle must carry its weight—along with the weight of its fuel—and achieve sufficient velocity to overcome the force of gravity. That means pushing the vehicle to a speed of at least 25,000 miles an hour to escape Earth's gravity.

If that vehicle travels to another body in space, such as Mars, then that body's own gravity well presents another challenge. The vehicle must generate enough force to reduce its speed to zero. Then, if a return trip is planned, the vehicle must burst out of that gravity well and, as it approaches Earth, decelerate.

All of these feats deal with a single physical challenge: to overcome the forces of gravity. Einstein described the force of gravity as a curvature of space; in other words, a well shape, with Earth at the bottom. Much of the efforts of commercial space companies deal with the problem of overcoming gravity enough to get satellites and vehicle into Earth orbit, and slowing down reusable rockets so that they return intact to the surface. To the scientists and engineers involved in these efforts, the Gravity Well is about gravity: a powerful force on all objects.

The Well *is* about gravity. But it's something else as well.

2. A REGION IN SPACE WITH ITS OWN TERRAIN

The Gravity Well does not look like a well when we consider it from our perspective here on Earth. It seems more like a wall than a well—or, rather, a mountain. This vertical peak has a few flat ledges and a few small valleys, all defined by gravitational forces. The flat places—Lagrangian points—are where the pull of bodies in space cancels each other out, allowing a spacecraft or base to remain stable relative to those bodies, without expending energy. The other side of the Gravity Well's mountain comprises a downhill slope, this one less steep, leading to Mars.

As the planets circle the Sun, this terrain changes with their relative positions. Unlike terrain on Earth, space terrain is dynamic. But it works the same way as the terrain we know, with hills and even valleys.

This region within the Gravity Well and just beyond it consists of the most rugged frontier we have ever encountered. Rough, hostile, yet filled with riches: precious minerals; energy; water and other means to sustaining life; as well as future scientific discoveries.

This is the new New World, encompassing low Earth orbit, middle orbit, geosynchronous orbit, the Moon, Mars, and a number of potentially valuable asteroids. Low Earth orbit is the frontier equivalent of the American East in the 17th century, with a small but fast-growing economy. Gradually, the farther regions are being explored and "settled," with public efforts—exploration, discovery, military efforts, and subsidization—gradually giving way to private investors and for-profit corporations.

The Gravity Well is where the next great economy will form, one with the potential eventually to dwarf the one on Earth. The people and nations who understand this best will be the ones who come to profit the most, in terms of the economy, technological prowess, and international leadership. America has a head start in technology and investment, but it has yet to meet its potential. We have other rivals—China, India, and Europe—who are eager to compete in this region.

We will conquer this frontier only if we recognize the third aspect of the Gravity Well.

3. A STEEP TRAJECTORY OF RISING ACHIEVEMENT

We stopped at the Moon, because the Moon was our goal. In the next great era of space, we must not pull back after achieving a single step. For Americans to support the space program in a way that brings the benefits I've described, we must see the overall mission not as a Moon mission, or a Mars mission, or an asteroid mission—but as a mission to ascend the Gravity Well. It's the space equivalent of the American West. We explore and we settle, and neither our setbacks nor—in the case of Apollo— our accomplishments can make us pull back.

The Gravity Well, in other words, presents a vision for our President, Congress, and other leaders to offer. We can share in that vision—and in the immediate benefits that come in the attempt. With each interim goal met, we gain the confidence to proceed still higher; because escaping the Well is our ultimate, most audacious goal.

Seeing the overall mission as one of conquering and then escaping the Gravity Well lets us focus also on the need to settle as well as explore. It is this connection that makes this investment so practically valuable for everyday citizens. After the scouts (the probes and astronauts) come the outposts (the human bases on a Lagrangian point, the Moon, and then Mars). Behind the outposts, entrepreneurs expand the transportation industry and high-tech manufacturing and even mining. As we rise within the Gravity Well, the economy—jobs, new products, and growth—follows. Private industry "settles" orbital space, then the flat terrain, and eventually the Moon, Mars, and the asteroids.

Each new achievement inspires young minds, wins the admiration of the world, and becomes a new baseline for growth. The beautiful thing about this perspective of the Gravity Well: the effort (technological, economic, and moral) reaps immediate benefits. The previous chapters defined them.

President Nixon had the right idea to weigh and prioritize our needs. The problem was, his calculus was dead wrong. From a system engineer's perspective, few other national needs can be more important. Because of the value delivered, calculable and incalculable, the Gravity Well must be an American imperative. It represents the single most important program for our future.

10

Mission Control

I was talking to a friend about America's civil space program recently, telling all about its benefits to the economy, STEM, and America's standing in the world. Just as I was winding up, he interrupted me.

"What if you were king of America? For a whole decade. What exactly would you do with space?"

The question threw me. Having worked so many years for NASA, I'm not used to thinking like a king. I like the idea of having a rational, practical, representative Congress writing laws and keeping an eye on the chief executive. I like the idea of giving all space stakeholders a say—including the scientific community, the aerospace industry, and every taxpayer who cares. Still, imagining yourself king makes for a useful thought experiment. If you were a benevolent king, with the intention of setting things aright in America and then restoring democracy after a decade, what would you do with all that power?

Personally, I would ban $4 cups of coffee.

Next, I would use space as a great tool to produce the out-sized benefits to the nation that we all want: a nonlinear effect.

Previously, I described society as a complex system, with stocks and flows, feedback loops, and degrees of resilience. Focus on just one part without thinking about the entire system, and you can unwittingly bring about sub-optimization: the tendency of one part to dominate the others, throwing the whole thing out of kilter. Complex systems tend to be nonlinear; one input, or one change to a part, can produce effects that far exceed the size of the input or the importance of the part.

But if I were king, why would I have to worry about systems thinking? Because if I'm actually to do some good during my decade-long rule, I need to fully consider our complex American system. Too often, our politicians and pundits use simple analogies to describe America and its government. We're a business. Cut the waste. Hand over as much as possible to the private sector. Compete with other nations. On the other hand, We're a family, the brotherhood and sisterhood of humanity. Or We're a village.

All of these analogies can be useful for devising solutions to our problems. Business in particular has created all sorts of innovations that have made government more efficient, communicative, and forward-looking. But society isn't a business. Nor is government. As we saw in Chapter Five, businesses take on calculated risks, but they reasonably shy away from uncertainty. Chapter Six recounted how NASA pioneered in technical innovations that helped seed the high-tech economy we have today. While a business is proprietary, protecting its knowledge, much of non-military government research creates a public knowledge resource: a renewable, minable ore of intelligence.

Not that society is government, or vice versa. Society is a system, and government is just one part. The private sector is another part, along with families and villages. The question isn't whether to favor one part over another. If we're using a systems approach—and, if I were king, we would—then I would look for the most positive nonlinear effects to the system.

Remember our lame car in Chapter Seven? A limited perspective leads to our buying an air conditioner and making the car go even more slowly. That's a negative nonlinear effect. If the car were a simple system, such as a stable room that simply needed cooling, then the air conditioner might have been the perfect solution. Instead, the car is a complex system; the best solution lay in a positive nonlinear solution, a relatively small change that would result in the biggest change for the good.

Back to my kingship. I would treat America the way I did missions at NASA. I learned you can't just focus on one subsystem, no matter how important, or the whole thing won't work. All the people in each area had to be systems thinkers, and they needed a chief systems thinker to integrate their inputs and make sure the interactions within the system didn't cause the system to fail.

The more complex a system becomes, the more important it is to think systematically. And America is enormously complex—almost as complex as your own body. I believe a systems approach works for America at least as well as it did for me at NASA. So, if my goals were to create the brightest future for the American system, increasing the odds of bringing the greatest happiness for the most Americans over the longest time, I would look for the tool or input with the greatest nonlinear effect: a small input with an enormous, beneficial output.

You can guess which tool I would choose.

There undoubtedly are tools or inputs other than space which would benefit America's future. But I challenge you to find one that has already been proven to work, and that can create the most benefits for the least cost. As we've seen throughout this book, the input of a tiny fraction of the federal budget into the public space program would boost the economy, restore America's innovativeness, and ensure our peaceful leadership among nations and restless young people around the world.

In sheer dollar calculations, the benefits over the next generation would total in the trillions. The cost: about $10 billion added to the annual NASA budget.

Here are some general directions for achieving these national benefits, maximizing the public space program's non-linear input to the American system. Yes, I am still king in this thought experiment. But the beautiful thing is, I don't have to be. This all can work without a king. Just keep in mind that each action is dependent on every other, so all must be part of meeting the challenge.

1. USE SPACE TO HELP SOLVE THE BIG PROBLEMS

Rather than operating on the competing demands of space constituencies, NASA and the White House should expect space programs to help tackle the biggest challenges facing the country. Just to name a few...

Mission to Planet Earth

Space offers an unparalleled vantage point for viewing Earth. Only by backing away from our planet can we truly see our tenuous life and how closely related we are. This perspective is critical if we are to understand enough about Earth to make good policy decisions on the ground. I've already described the space-based measurements that revolutionized weather prediction. That is just the tip of the Earth system science iceberg (to use an unfortunate analogy). Only from space can we obtain the global data we need to unravel the mysteries of our home planet.

For example, it was satellite data that discovered the ozone hole in the 1980s. Satellite data are currently showing us how fast the Arctic sea ice is melting. We monitor pollution globally to improve air quality around our urban areas. And we are developing better knowledge of how the climate works, thanks to satellite measurements showing how much sunlight penetrates the clouds and hits Earth, and how much heat leaves it. Clouds may hold the clue to how to manage climate change in the coming century.

Clearly, Mission to Planet Earth is more than a science program. It provides the knowledge we need to make decisions, ranging from international policy to national laws, from state and local planning to business strategies. This mission even helps you and me decide how we will live—and what footprint we will leave for future generations.

Our Next Economy

Since the turn of the century, political leaders and prominent economists have described the problem of America's economy as a loss of manufacturing jobs. This conclusion is understandable. The last century's economy built itself off a growing middle class through well-paid manufacturing jobs. The middle class in turn could afford to buy the marvelous products of those factories—cars, televisions, washing machines—and they flew with the modern new airplanes in previously unimaginable numbers.

Yet that economy no longer drives America's prosperity; manufacturing knowledge is now widely available. We have been transitioning to a new economy, a sea change that rivals the last century's transition from an agricultural to an industrial society. Today we find ourselves competing with many other nations. Manufacturing expertise, once the purview of so-called developed nations, is now widespread—particularly in Asia and South America. To regain our competitive leadership, we need to build our economy on something new, using a quality and degree of expertise no other nation has obtained. What could that new economic engine be? And are we capable of creating it?

We have a brilliant precedent: the flying machine.

Just as we built the aircraft industry and the air transportation market that literally rose from it, we can do the same with low-cost space transportation. I have no doubt that it will be a driving force for the rest of this century. Inexpensive, reusable space vehicles will carry unprecedented amounts of cargo into low Earth orbit; carry passengers from one continent to the next in under two hours; build manufacturing facilities and commercial laboratories in microgravity environments; and, eventually, mine the treasures of asteroids beyond Mars. While some of this may sound far-fetched, imagine how someone in 1900 would respond to predictions of low-cost air travel, of air cargo, of the Boeing 747 and its cocktail lounges in the sky. Already, investors are betting billions in private capital on an equally brilliant space future. America has a natural advantage—and a big head start. To foster the next great, tech-driven economy, government and the private sector must work together to conquer the Gravity Well, with their combined efforts focused by our civil space program.

America's Imperiled Reputation

Over the last 100 years, brave Americans on the battlefield and industrious Americans back home played a major role in making the world's citizens more free, secure, and prosperous. For most of that time, most of the world viewed us as—mostly—a force for good. Now we have another battle to fight. It's one in which the enemy has new psychological advantages. Despots and fanatics paint a picture of America as intent on exploiting others for its own power and wealth. Our answer has been mostly military, with boots on the ground and drone strikes from the air. All with good reason. Yet we must not forget the wisdom of President Kennedy, who understood that we needed to win both the war and the argument. Persuasion, through demonstration of universally valued attributes, must supplement our military force. Kennedy knew that he couldn't impose our capitalist system on other nations. He envisioned the Moon shots as a way to prove the superiority of the American Way without using force.

The secret of Apollo: we shared it. Our President announced to the world a seemingly impossible goal, and then we openly communicated both our progress and our setbacks with people around the world. When we succeeded, we won more than a race against the Soviets. We won the hearts of countless individuals and national leaders, globally, without firing a gun.

This powerful tool of international influence is still available to us, if we choose to use it. But it must be awe-inspiring. We'll get to that in a bit.

The Scarcity of American STEM Students

Chapter Seven showed the scope of the problem while demonstrating how NASA can suck bright young minds into space. If I were king, I would set our nation's space program on a trajectory that would inspire not just the international community but our own youth. Using the Gravity Well as a focusing challenge, we could, one step at a time, explore and settle the terrain of space in our neighborhood of the Solar System. Today we are well on the way toward a developed economy in low Earth orbit. Farther out, NASA missions can deliver outposts at the Lagrangian points, the Moon, Mars, and asteroids, all within the next generation of students. But only if we set our American know-how to the task. A full effort will guarantee a dramatic uptick in the number of students in math, engineering, science and technology students. And it will nurture highly paid American workers who will drive our economy well into the future. In short, the drawing power of space must be utilized strategically.

Our Crumbling Transportation System

Our economy flows through efficient, clean, low-cost transportation. At the same time, we must reduce our use of fossil fuels if we want to limit climate change. NASA's aeronautics technology can serve as the key to this effort by developing cleaner, safer, more efficient aircraft. The next generation air traffic control system is well within reach of near-term deployment. The FAA and NASA will have to work together with industry representatives; and research and contract funds need to be freed up to bring the system online.

Sometime in this century, people will be flying autonomous aircraft—the equivalent of the self-driving car. Personal air vehicles, perhaps powered by electric batteries, can help relieve our overcrowded airports, save land, and minimize emissions. Yet, over the past 30 years, budget cuts have hobbled NASA's aeronautics mission. I would reinvigorate that mission to help create the next economy.

The Need for a Forward-Looking American Vision

To maximize the space effect on all these problems, I would bring together teams consisting of key Cabinet members—State, Labor, Transportation, Commerce, and others. The goal should be to innovate solutions using space exploration, science from space, and technology for space. Our leaders must not view NASA merely as an exploration agency or a tech agency. It is both of these, of course. But in addition, the agency can serve as an essential tool to solve our problems.

Again, think of space as a sub-element in the complex American system, capable of producing an outsized effect. An American leader who takes this systems view will employ the brightest minds inside and out of government to craft a new vision for our nation—one based on the best we have to offer.

2. FUND NASA'S MANDATED MISSIONS

Over the previous three decades, our nation's leaders have reduced NASA's buying power by 25%. This decline has spurred some destructive infighting among aerospace stakeholders. Private industry attacks NASA, saying the bureaucracy is too slow and risk averse. Universities attack NASA because their grant programs are getting cut. Scientists attack aeronautics, space technology, and human exploration because the money could be diverted to science missions.

If I were king, I would work to get everyone into sync. The academic community, industry, and NASA all must work together to move our programs forward. Exploration without science is shortsighted. Science without exploration engineering

179

leads to a dead end. In order to gain understanding of Earth, we must get off it. To understand the Solar System, we must get out into it. Science and exploration will stall without technology, including aeronautics. Costs will rise, capability will shrink, and our leaders will see fit to reduce the public space program still further. The only choice for the space community is to work together.

I'm not being a Pollyanna here. One of the most politicized communities, academia, already does a good job at cooperating. Every ten years, the National Academy of Sciences (NAS) brings the science community together to look at what we know already; and this group determines the highest-payoff investigations—what we need to know next, to know the most. The NAS gives that list to NASA, providing advice for setting the priorities for research in space.

We don't have a similar method for human exploration—of polling experts to develop a decadal set of advice. One reason is that human space exploration is a relatively immature activity. Science has had a 500-year head start. While universities are set up to do astrophysics, with a department in most universities, that's not true of human exploration. While I would stop short of establishing a human exploration department within universities, I would recruit the physics, chemistry, biology, and aerospace faculties, financial and business departments, and other disciplines. Together they can help develop the human exploration priorities in a reinvigorated and properly funded program.

On the other hand, we could simply limp along on NASA's current budget and shift more funding over to science programs. We could go to Mars with robots alone and still conduct good science. But that approach will never get us to Mars; and we happen to be humans, not robots. Imagine if we had only explored the American West using robots! Eliminating human space flight would abandon many of our most ambitious goals, including the establishment of extraterrestrial colonies, mineral extraction, and manufacturing.

Then again, we can't abandon science for human space flight. In order to allow for human exploration on Mars, for example, we need to understand more about the property of its soils. Scientists have already discovered that the Martian surface contains a great many carcinogens. (Sorry, Matt Damon. Those potatoes might not have been as good for you as you thought.) We need more data to figure out how to make a spacesuit that can protect astronauts—and settlers.

The same goes for missions to asteroids. Human exploration and science need to work together if the space program is to serve the entire American system. Asteroid missions can work the way Apollo did. On the last several Apollo missions, the science community worked with the astronauts for many months to maximize the knowledge from human flights. The Apollo astronauts picked up many rocks on the Moon, and scientists are still getting data from them. We know much more about the history of the Solar System, Earth, and the Moon because of those rocks. Could they have been collected by remote vehicles? Certainly. But not as well. Human eyes, knowledge, and intelligence are far more capable of negotiating unknown environments than machines are. Besides the knowledge ore extracted from rocks, there is the knowledge that came from putting humans on a foreign body. That intelligence improves our lives on Earth and will help put us on Mars.

The answer to getting all these needs in sync lies in the budget. If you have the budget you need, it is much easier to put cooperative programs in place. A fully funded space program offers the single best answer to all of its stakeholders working together.

In relative terms, it's not an expensive answer. If you were to create a normal-sized pie chart of the overall federal budget, NASA would be invisible. Suppose you doubled NASA's budget. It would still fail to show up on that chart. And I am not suggesting a 200% increase. I'm proposing bumping up NASA's budget by half.

We've seen the benefits of funding NASA's missions. During the Sixties, the space program jumpstarted the satellite and weather-prediction industries; these benefits more than paid for the agency's entire budget. In fact, the tax revenues from those industries alone are enough to pay for NASA's current budget.

But what we saw in the Sixties is not happening today. The current NASA is pulled in different directions by Congress and the White House. Both branches have excellent reasons for their policies, though those reasons are not well articulated for the public. The problem is that the budget has not allowed NASA to go in both directions at once.

The problem started during the George W. Bush administration, though the White House itself was not to blame. Bush's NASA chief, Sean O'Keefe, came into office with some serious problems to solve. The International Space Station had a cost overrun of billions. O'Keefe brought an accounting eye to the agency; in a previous job at the Defense Department, he'd been known as the "Grim Reaper." Several years after cutting costs at NASA, O'Keefe received another directive from the President: restructure further to prioritize a mission to Mars.

In 2004, President Bush himself came to headquarters to announce the mission. This was the only time a president came to NASA to speak to us during my entire career at the agency. Ever since President Nixon cancelled the Apollo Program, condemning us to the bottom of the Gravity Well, NASA has had to lower its sights for human exploration. With Bush, it looked to us at NASA that we were finally seeing a positive change in direction. Here was a president willing to take us beyond the Well, matching the ambition that the last Texan president had shown for space. The Bush policy sought to achieve NASA's potential—inspiring the nation, pulling young people into STEM, and enhancing America's international reputation. Congress, in a bipartisan move, actually authorized Bush's vision, but the budget never caught up. Congress failed to appropriate the money to match the mission. Unfortunately, we never saw Bush at NASA again. In any case, his program, though well thought out, was not part of a broader national policy. Nor did he properly explain the Mars mission to the wider public, or make much of an attempt to garner its support.

Sean O'Keefe's successor, Michael Griffin—an engineer and NASA veteran—tried to shift the existing budget to focus on getting to the Moon, the first step toward Mars. He took flak from the scientific community for pulling funds from the science mission to pay for human flight.

Then along came a new administration with different ideas.

President Barack Obama aimed to foster a new market for space transportation, just as NACA helped birth air transportation a century before. The new deputy administrator, Lori Garver, had previously worked for NASA in policy positions before joining a space lobbying group. She was a passionate advocate for commercial spaceflight, and brought that view with her when she joined Obama's transition team before the inauguration. Technology and growth of a commercial space industry became the priorities. The goal was not just to enrich government contractors but to form a real market.

Despite this new emphasis, NASA continued to promise a Moon base by 2024. The Obama vision didn't replace the Bush vision; the Republican Congress assured that wouldn't happen. Instead, the new ambitions were simply added onto the old ones, without the budget to go with them.

That changed in 2010, when Obama cancelled the Moon base mission, announcing instead that NASA would send a mission to a nearby asteroid by 2025. The problem was that the deadline was nearly impossible; a spacecraft could not physically reach any suitable asteroid by that time.

In short, Congress handed NASA one set of priorities, and the administration gave it another. The President's priority on the commercial crew program offered a national benefit different from the congressional priority—to build a massive rocket and new spacecraft. The White House's commercial approach would create the next market and catalyze the next economy. Congress's big-rocket approach would let us lead the world into space, while assuming incalculable risks that no company, or any other country, would take.

The choice never got made. Congress and the White House simply acted as if it were possible to get both the international leadership and the economic growth benefits without paying for both. Because the budget did not increase, the agency came under unprecedented criticism for failing to manage it properly. The fighting outside the agency resulted in wasted internal efforts. Schedules slipped, morale dropped. To critics of government, NASA seemed a prime example of bureaucrats gone wild.

Which would you choose: international influence through pre-eminence in space, or the next new economic market and the associated industries, companies, and jobs?

Now here's a better question: why should we have to choose? Aren't both worth the money? The previous chapters have sought to prove that they are. We want the leadership for our security, and we want to catalyze the commercial space program for our future prosperity. Does security outrank prosperity? Or would you choose the economy over safety? To my mind, both remain top national priorities. We should not rob Peter to pay Paul. To say we have to give one up in order to keep NASA's budget level: that's a false choice.

If we agree to keep all the programs with sufficient funding, then the next step is to build a program across those visions so that they mutually support each other. We need science to feed the knowledge ore and advance humanity. We need human exploration to inspire the world, and to create options to settle new worlds. We need aeronautics to help create the next aircraft, rockets, and control systems. And we need space technology to catalyze the next space economy. Fund them appropriately, and their sum is far greater than their parts.

And don't just fund them over one congressional term, or one administration. We must fund them for a generation. If I were king, I would increase the overall NASA budget by about $1.2 billion each year for eight years and then sustain that level of investment.

3. BUDGET MORE SENSIBLY

The most efficient way to execute a project is to undertake an in-depth planning study followed by a design phase with rigorous reviews, then go through a build-and-test phase and, finally, flight. This process demands a certain funding profile to shorten the schedule and minimize total costs. The common way of federal budgeting, a "flat-funding" profile that maintains the same amount of money throughout a project's lifespan, is inefficient.

What's worse—and all too common—is leaders' practice of starting, stopping, and restarting programs. That's more than inefficient; it's wasteful in the extreme.

Projects often suffer from both bad budgeting habits: flat funding and starting-and-stopping. Work gets going with an even budget, Congress changes the budget, and the project grinds to a halt. For example, some work got stopped on key subsystems for Orion, the deep space human spacecraft, in order to let the big-rocket Space Launch System catch up. Because of flat funding, NASA couldn't work on both at once.

Similarly, key Earth science missions get stretched out in wasteful study phases that keep a core team together but without the funding to move forward smartly. The launch date for the high-priority Mars sample return mission, which will bring a piece of Mars back to Earth for laboratory analysis, has remained ten years away for over a generation now.

The alternative to flat funding is the deliberate injection of additional funds as a project meets key targets. This profile becomes feasible when Congress funds NASA with a steady increase over the next eight years.

4. RENEW NASA ITSELF

This one is hugely important for mission success. The agency has been suffering for two decades from the effects of American indecision about its space program. Declining spending power has resulted in an aging infrastructure and, to some extent, a brain drain. Some of the most talented, experienced scientists and engineers are retiring or simply leaving. As with any large middle-aged organization, NASA needs some revitalization. This is not a reflection of NASA's scientists' and engineers' ability or commitment. I know a great many of them, and, given the circumstances, their dedication has been amazing.

The good news: current leadership is working on renewing the agency.

The bad news: they're working without the audacious national mission we need. It's as if NASA is a boxer training without any fight scheduled.

I have hope that the next president and Congress will craft just such a mission. When they do, though, they must include a study of how to upgrade NASA to meet the challenge. That includes improving the ways the agency interacts with other parts of the executive branch.

Just as importantly, space must be de-politicized. The space program already has citizens' widespread bipartisan support. There's no reason for the political parties to carry out their ideological agendas with space. Solving our nation's problems, spurring the economy, inspiring STEM, leading the world: All Americans want these things. As the next chapter will argue, space is too big for provincial politics. And the space program itself represents the best expression of the American destiny.

5. INSPIRE THE PUBLIC WITH AN AUDACIOUS MISSION

To realize the national benefits I've described will require a degree of presidential and congressional leadership that we have not seen in a long time. More importantly, space must have the backing of the rest of us. Each of us needs to understand the broad benefits. And we must demand that our leadership ensure we reap the rewards of having the most advanced space-faring technologies in the world.

For any of this—the leadership and the support—to happen, first we need a vision. We need a mission that strains our capabilities, one that's hard. NASA hard. The ideal mission will have certain characteristics:

It will seem crazy to skeptics, while stretching our belief in ourselves. Just as Kennedy's call for going to the Moon, this new mission must thrust right up to the edge of impossibility.

It will be nonpartisan and represent no particular ideology, except the characteristic American quest for invention and discovery.

It will be new, exceeding all the goals of previous presidents. Bigger and newer than Lyndon Johnson's vision of a Moon base. More daring even than George W. Bush's call for a base on Mars. Bolder than Barack Obama's plan for capturing an asteroid.

It will deliver—immediately and in the long run—all the economic, STEM, and international benefits.

What in heaven could such a mission be? I've been suggesting it throughout this book. It's the Gravity Well.

The mission itself will not be enough. The next president should enshrine that mission within a vision, one that describes the best possible future for America and the planet. This vision, too, must be bold enough to stretch credulity while captivating our nation's most restless young intellects. It could go something like this:

> "Our generation is about to take the first step toward turning
> Earth into a homeland that guarantees all individuals the right
> to reach their potential. We will achieve this feat by stepping
> up and away from Earth."

The President will then describe the Gravity Well, the definable part of the near Solar System that will serve as our next, greatest frontier.

> "Conquering this Well means exploring it, then settling it—much
> the way we explored and settled the vast American frontier.
> And in doing that, we will find riches we can only begin to
> imagine.
>
> "We will manufacture new machines, and create new gener-
> ations of computers fully capable of artificial intelligence. We
> will invent new smart materials, with clothing that changes
> temperature with the weather, and self-healing polymers. We
> will have new medicines, developed in microgravity environ-
> ments. We will tap into limitless supplies of energy without
> causing any harm to our planet. Future generations may even
> grow enough food in space to satisfy everyone. One day,
> our children's children will turn vast tracts of Earth into parks,
> because they will have all the terrain and all the space they
> need, off this planet.
>
> "This vision sounds far-fetched, I know. But so did President
> John F. Kennedy's call for landing people on the Moon and

bringing them back safely. When he made that call in 1961, commercial airplane flight was barely 50 years old. Only a tiny fraction of Americans had even flown in an airplane. And who would have thought that our giant leap into space would come with so many added benefits, like accurate weather prediction, GPS, and our smartphones that communicate with satellites?

"It has been more than half a century since President Kennedy made that call. And now it is time for a new vision, one that reaps even greater benefits.

"Ten years from now, we will have a base on the Moon. Twenty years from now, astronauts will be operating on a permanent base on Mars. Twenty-five years from now, we will be mining the asteroids near and beyond Mars. A single asteroid, no wider than your living room, can contain $10 billion worth of gold, along with platinum, tungsten, and the rare earth metals we desperately need here, where supplies are running low. Along with those metals, we'll extract iron, nickel, aluminum and titanium to construct new bases on Mars and in space. On other asteroids we'll obtain water and oxygen to sustain life, as well as hydrogen, oxygen, and ammonia for fuel.

"Beyond those 25 years, we will create an economy in the Gravity Well that dwarfs the one on our planet. We will achieve a global prosperity that we cannot fully imagine today—just as no one could envision the wealth we have achieved since the Wright brothers first took to the sky.

"But first, we must step up and climb to the top of the Well. If we do that—if we meet that challenge—then we will be known as the generation who opened up the terrain of space, who settled the newest, greatest frontier. Future generations will remember us for our courage and our boldness. And they will remember us for something even more important: That we did it peacefully."

11

Vault of Heaven

Up until fairly recently in human history, people believed that a solid dome, studded with the Sun, Moon, and stars, lay above Earth. The ancients called this dome the firmament. Above it, the ancient Jews taught, lay the vault of heaven. Our human instinct—something in our DNA, perhaps—compels us to rise beyond such barriers. The aerospace equivalent of the firmament is the Gravity Well. We live more or less comfortably within it, a million miles below its star-studded top.

Our ambitions and technology, our human drive, have lifted us above the surface of Earth to perform amazing feats—all within the bottom eighth of the Gravity Well, with humans venturing less than a thousandth of the distance from Earth to Mars. In order to go the rest of the distance, colonizing the next planet and preparing for a future unfettered by the Gravity Well, we Americans must repeat what we did over the previous century.

First, we fostered a society that encouraged inventiveness and entrepreneurship.

Next, we catalyzed the growth of an aeronautics industry through research and lucrative contracts (U.S. mail, the War Department, NACA).

Once aviation moved beyond the realm of economic uncertainty to a calculable risk, we—society, the federal government—got out of the way as much as possible. We deregulated the airlines while ensuring the safety of air traffic, maintaining aeronautics research.

Then we—society, the government—moved to the next level, literally: space. Again, our whole nation bore the burden of risk and danger, the uncertainty. Together we mourned and honored the sacrifice of the astronauts who gave their lives to the venture. Together, the whole world celebrated when Neil Armstrong took our first step on the Moon.

As we moved beyond the nearer frontier of low Earth orbit, industry began to move in, building satellites and launch vehicles more quickly and cheaply with the advantage of multiple production. Uncertainty in LEO is moderating into economic risk, making near space a ripe field for entrepreneurship.

This book has shown that private space could not happen—and would not be happening now—without vigorous, sustained public effort. Even today, Elon Musk's SpaceX depends on NASA funding. I hope I have shown that, while we celebrate the achievements of Musk and his fellow entrepreneurs, we remember that we have barely begun as a society in space. Hardly a day goes by without news about SpaceX's latest amazing barge landing, or the James Webb telescope's progress, or an unprecedented view of Jupiter and its moons, or Hollywood's next space blockbuster. Americans are clearly excited about space, and believe in a space future. Most citizens have a positive attitude about space, including the public space program.

Yet some confusion remains about what, exactly, space entails. The space of the ISS, communications satellites and planned tourist flights lies near the bottom of the Gravity Well. While Musk promises to help lead an effort to escape the Well, he is the first to say that it won't happen without NASA.

Meanwhile, Americans seem to find their country stuck in place, with a deadlocked Congress and a deeply divided electorate. Spiritually, our nation lies stuck at the bottom of the Well, uncertain of how to break free. And yet, this is the beautiful thing about the space program: The Gravity Well is not just a metaphor.

At the same time, escaping the Well—even the attempt at an escape—will break us out of the moral well we lie in today. By catalyzing our economy, STEM, and our standing in the world, and helping ensure the survival of our species, the Well becomes more than a barrier. It leads us to... what? Call it the vault of heaven if you like. Or call it the future.

Above the Well lies Mars and the universe beyond. As we rise out of the Well, we create a renewed economy, inspire inventors and discoverers, and foster a more cooperative Earth.

This rising should come naturally to us. Thousands of years ago, people made a sustained, dangerous effort to cross the land bridge to America. This journey, and the waves of settlement in the New World, doubtless entailed both individual inventiveness and the backing of entire restless peoples.

We Americans are the epitome of that individual drive and societal restlessness. It's where our greatness lies. The moment we refuse to move; when we find comfort in the low places; when we fail to leave our home fires and face the dangers, to develop the systems and the technology—the moment we stop, we stop being fully American. For that matter, we stop being human. Our civilizations, built on growth and new ideas, begin to decay. America begins to decay. You and I decay.

But I don't believe that will happen. The Gravity Well presents today's greatest exploratory challenge, and the single best direction to move our society forward. If we agree as a people, a government, and a world to move up and beyond the Gravity Well, then the higher we go, the less it will pull us back. Gravity will gradually release its hold, and we will loose the bonds that hold us below the firmament.

And then the well becomes a ladder.

Appendix I

What You Can Do

EVERYONE

Become a thought leader on space progress and issues.

Go to **TheGravityWell.org** for links to the best social media on space and space policy. Be sure and subscribe to the site's blog updates; and subscribe to one or more of the other sites listed. In addition, you'll find a list of organizations that focus on space. Consider joining one.

See Appendix II for talking points on the value of a public space program.

Join the conversation.

If you join the Gravity Well site, you'll be invited to comment on the blog posts. Also see the Facebook page, **The Gravity Well**. Twitter: **@thegravitywell**.

Of course, not all conversations have to be online. Make sure your circle of co-workers, family, and friends see space as a priority, and invite them to join the discussion.

Push your Congress Members.

On **Congress.gov** you can find the contact information and voting records of your House Members and Senators. The House of Representatives recently re-launched the defunct space power caucus, a bipartisan group that focuses on space's role in national security. It would be helpful if the House had a caucus to promote space for its economic and STEM value as well. Ask your member to join the space power caucus and urge the formation of a civil space caucus. There used to be one, but it hasn't been robust since 1989.

Both houses have an aerospace caucus, which covers a broad range of issues involving civil aviation, defense and national security and space. See if your senators and representative are on this caucus, and urge them to form a separate space caucus.

If they don't seem to show any interest in space, urge them to pay attention to its value for the nation. (And don't be afraid to send them copies of this book!)

EDUCATORS

Invite me to give a talk.

Get in touch with me through **TheGravityWell.org**. I'd be happy to arrange a Skype session with your class. And I might be available to speak to your institution.

Take advantage of the significant educational material provided by NASA.

Contact a NASA Center (see Appendix VI) and ask for help. Each center has resources specifically designed for educators.

Here are a few of the most notable programs:

Human Exploration Rover Challenge   High school and college teams of six must design, build, and race a human-powered rover through an obstacle course designed to simulate the terrain of rocky planets like Mars.

NASA Student Launch Teams design and launch high-powered rockets to study propulsion systems. Varying challenges are offered to students from middle school through graduate school, as well as non-academic teams.

The FIRST Robotics Competition High school teams of 25 or more students have six weeks to build and program a robot to complete challenging tasks against competitors. NASA funds this.

Sally Ride Center EarthKAM In this NASA educational outreach program, middle school students can request images of specific locations on Earth taken from space. Sally Ride, the first woman in space, founded this project in 1995.

Appendix II

Is Space Worth It?

The most common argument against a publicly funded space program is that we can't afford it, given the government's many other priorities. The most common answer is that space makes an excellent public investment.

But if it is a good investment, why not leave space to private investors? Chapter Five offered economist Mariana Mazzucato's argument that publicly funded missions help turn economic uncertainty into calculable risk. By her definition, calculating the return on public space expenditures becomes impossible.

On the other hand, if you look at the American economy before, during, and after Apollo, you can see a correlation between economic growth and a robust aerospace program. During the Fifties, the American gross domestic product grew by 26% (from $2.27 trillion to $3.06 trillion in inflation-adjusted 2009 dollars, according to the U.S. Bureau of Economic Analysis.) During the Sixties—the decade of Apollo—the GDP grew by 35% ($3.08 trillion to $4.72 trillion). During the 1970s, the GDP grew by 28% ($4.71 trillion to $6.5 trillion).

1950-1959: 26%
1960-1969: 35%
1970-1979: 28%

In other words, the American economy grew one-third faster during the Apollo decade than during the decades before and after. Of course, no economist would say that the space program *caused* this growth. Still, as Chapter Six showed, the "knowledge ore" of science and technology generated by the Apollo program and NASA aerospace R&D continues to provide spinoffs today.

Total cost of Apollo in current dollars: $109 billion, or 0.2% of the GDP, a slice equalling one five-hundredth. Were the economic benefits worth that cost? You decide.

Besides making an economic case for a fully funded public space program, Chapter Seven argued that space inspires young people to go into STEM (science, technology, engineering, and mathematics).

> Annual cost of American education: $810 billion.
> Approximate cost of STEM: $400 billion.
> Additional cost of expanding STEM expenditures by one-half: $200 billion.
> Additional cost of expanding NASA's budget by one-half: $10 billion.

Chapter Eight made the case for space's value in enhancing America's international standing. Assuming that the value of something is what you are willing to pay for it, we can look at relative costs.

> Annual federal expenditures on international affairs (State Dept., Defense, Energy, Veterans Affairs): $800 billion.
> Annual NASA budget: $18 billion.

Finally, we can look at how much of an impact the public space program has on the federal budget.

> Size of 2015 federal budget: $3.7 trillion
> NASA's share of the federal budget: 0.4%
> Impact to the federal budget of fully funding NASA: 0.3%

Still, some people say the money would be better spent on feeding and housing the poor, or on medical research, or on non-space research and technology. In fact, the federal government spends far more on these programs already.

Annual federal housing assistance: $190 billion, according to the Center on Budget and Policy Priorities.

Annual budget of the Supplemental Nutrition Assistance Program (SNAP): $75 billion (same source).

National Institutes of Health budget: $31 billion, according to the NIH.

Federal R&D expenditures: $145 billion, according to the American Association for the Advancement of Science.

Can we afford space? The numbers seem to show that we can; the impact of fully funding NASA to meet all of its congressionally mandated missions is minimal—almost undetectable in the larger scheme. But do these numbers prove the value of space? What is the total worth of building a vast new economy in the Gravity Well? Of providing future generations with the next inventions and discoveries? Of proving to the world that America can lead peacefully? Of ensuring the long-term survival of humanity?

These benefits, I believe, are priceless.

Appendix III

Recommended Books and Online Resources

BOOKS

William E. Burrows, *The New Ocean: The Story of the First Space Age.* At once entertaining and thorough, Burrows's book guides us through the beginnings of space up through 1998. *The New Ocean* was a finalist for the Pulitzer Prize in 1999. In my prejudiced opinion, it deserved to win.

James Hansen, *Engineer in Charge: A History of the Langley Aeronautical Laboratory, 1917-1958.* Hansen, a history professor at Auburn University, is the author of the bestselling book, *First Man*, the authorized biography of Neal Armstrong. *Engineer in Charge* covers the history of NACA (NASA's predecessor) and the early days of NASA, with a focus on technical innovations. The book deeply influenced me.

Michael J. Neufeld, *Von Braun: Dreamer of Space, Engineer of War.* The best biography of Wernher von Braun, the Third Reich aerospace engineer who helped found the American space program. An SS member and, arguably, a war criminal, von Braun was also perhaps the greatest space visionary.

Robert Seamans, Jr., *Project Apollo: The Tough Decisions.* Seamans worked at the highest levels in NACA and NASA from 1948 through 1968. In this monograph (available on Amazon), he offers an insider's view of the politics and technical challenges—not just of Apollo but of the Mercury and Gemini projects that preceded it. Seamans was more than a brilliant administrator and aeronautics engineer; he also proved to be an excellent true storyteller.

Neil DeGrasse Tyson, *Space Chronicles: Facing the Ultimate Frontier.* Space's closest thing to a rock star, Tyson has published essays in a variety of popular publications and online blogs. This book pulls together his best writing. Behind the wit and provocative storytelling lies an urgent message: America must renew our commitment to space before it's too late.

Chris Impey, *Beyond: Our Future in Space.* This distinguished astronomer describes the years to come as we explore space. Behind his detailed account of the future lies his belief (one I fervently share) that exploration is the compelling force of humankind's progress.

Lewis D. Solomon, *The Privatization of Space Exploration: Business, Technology, Law, and Policy.* Solomon, a lawyer, serves as a cheerleader for commercial space missions.

David S. Landes, *The Wealth and Poverty of Nations: Why Some Are so Rich and Some Are so Poor.* What makes a country rich and great? This engaging book is full of surprising answers. Besides luck (including the right resources), a wealthy nation also depends on systematic decisions by wise leaders. This book is where I found the story of the Chinese burning their beautiful ships and its effect on China's progress—the equivalent of America scrapping the Saturn V rocket.

Daron Acemoglu and James Robinson, *Why Nations Fail: The Origins of Power, Prosperity, and Poverty.* The authors show that political and economic decisions lie behind the huge gap in the growth and power of nations. Don't miss the section on America, which may be making the wrong decisions right now.

Mariana Mazzucato, *The Entrepreneurial State: Debunking Public vs. Private Sector Myths.* Mazzucato argues eloquently, with plenty of examples and data, that government is an essential catalyst to the private-sector economy. Among her most compelling examples of government success in stirring entrepreneurship: NASA.

M.I. Finley, *The Ancient Economy.* Written by one of the great modern historians of ancient Greece and Rome, this book—a compilation of lectures—shows how the richest citizens of Athens and Rome acquired their wealth by taking it from others. Before the modern, government-assisted economy and currency systems, people didn't get rich from growing the economy. Wealth in those days, as one reviewer puts it, was a "zero-sum game."

Tomas Sedlacek, *The Economics of Good and Evil: The Quest for Economic Meaning from Gilgamesh to Wall Street.* We're used to thinking of economies as states of growth and non-growth. Growing economies are good, and shrinking economies are bad. There's a lot to be said for that attitude. But Sedlacek, an economist at the University of Prague, argues for a more sophisticated, ethically centered approach. Economies, he says, can be a force for good beyond growth. Anyone looking for a thought-provoking way to examine the value of the space program should read this book.

Philip Kitchner, *Science, Truth and Democracy.* Is science intrinsically good and necessary, or does it merely serve the powerful? Neither, says this clear-thinking philosopher.

He advocates a public consensus on the scientific avenues to explore. Kitchner's book struck a chord with me. A vast majority of Americans see the value of space (if not all its many benefits). They should have a greater voice in spurring the exploration of space.

Stephen P. Johnson, *The Secret of Apollo: Systems Management in American and European Space Programs.* This book makes a convincing case that the systems approach to management was NASA's single greatest innovation. Besides serving as a scholarly history of the space program, this fairly slim book offers a fine introduction to anyone interested in systems.

Donella H. Meadows, *Thinking in Systems: A Primer.* The single best beginner's guide to systems, written by an environmental scientist. Meadows worked with a brilliant team at MIT to develop the new field of systems dynamics.

John D. Sterman, *Business Dynamics: Systems Thinking and Modeling for a Complex World.* A fat textbook, complete with CD-ROM, offering a course in systems. Sterman, a professor at MIT's Sloan School of Management, has greatly influenced my own thinking about space. Once you adopt a systems view, you realize that the space program is about more than space. It can lead our nation into its next great era.

WEBSITES

AllianceforSpaceDevelopment.org The Alliance for Space Development advocates settlements in space.

AmazingSpace.org The best astronomy site on the Web. If you're interested in the Hubble or James Webb Telescopes, here is where you'll find the latest data and images.

ESA.int ESA is the European Space Agency. The site contains the latest news from across the pond.

Lunar.xprize.org Google's $30 million X Prize spurs innovation in low-cost robotic commercial space exploration. As you might expect from Google, this site is very, very cool, including a documentary series produced by J.J. Abrams of *Star Wars* and *Star Trek* fame.

NASA.gov One of the least stuffy government websites, packed with multimedia. This is the place to learn how to participate directly with space, from getting your Scout troop's experiment onto the International Space Station to visiting the Johnson Space Center.

NSS.org The site of the National Space Society, which advocates human colonization of space.

Planetary.org The site for the Planetary Society, Mars Society, SETI Institute, Commercial Spaceflight Federation, and the Coalition for Deep Space Exploration.

Space.com   A great source of information, gathered by a commercial online publisher—news of the latest commercial and government launches, astronomy facts, and even an area where you can shop for a telescope.

SpaceDaily.com   The go-to site for space news. While geared toward the true space fan, space works on the brain a lot the way baseball does. Follow the players and the games (or, in this case, missions) for a while, and you will find yourself becoming a true fanatic.

SpaceFoundation.org   The Space Foundation leads educational programs and holds an annual symposium of space stakeholders.

SpacePlace.nasa.gov   A fun site for kids, with everything space and NASA.

SpaceWeather.com   Space weather affects telecommunications on Earth, and it can affect the timing of launches. For you and me, it's a cool way to see how the Sun is impacting (literally) Earth with its cosmic rays. A typical "weather conditions" announcement reads: "Solar wind speed 492.0 km/sec, three sunspots, none of them threatening."

SpaceX.com   Keep up with Elon Musk's ventures in space.

Stsci.edu   Site of the Space Telescope Science Institute, located within Johns Hopkins University. Get a close look at missions run by this inside player.

APPS

NASA (free)   Get the latest agency news, find the NASA facility nearest you, and catch a seemingly infinite number of videos and images.

Space, Astronomy & NASA News (free)   NASA's one-stop app for the latest space news.

Space Images (free)   This picture-packed app lets you share the most amazing images on social media.

New Horizons (free)   Explore Pluto and the outer Solar System through images from NASA's probe.

ISS Live (free)   This app makes you feel a part of the International Space Station, with virtual 3D tours and live stream of data.

ISS Spotter (free)   Find out when the International Space Station will be visible from your roof or backyard, then set an alarm to make sure you can see it. The built-in compass will help your spotting.

NASA Be A Martian (free)   This app lets you participate in the current (and future!) missions on Mars, with the latest images and news.

Rocket Science 101   This app, created by NASA's Launch Services Program, lets you build your own rocket for your favorite NASA mission.

Earth Now (free)   Another app from the Jet Propulsion Laboratory. This one uses maps to visualize recent data from Earth Science satellites, including surface air temperature, carbon dioxide, carbon monoxide, ozone, and water vapor as well as gravity and sea level variations.

Images of Change (free)   This NASA app lets you see just how Earth is changing, including then-and-now comparisons of glaciers, wildfire sites, and floods.

NASA 365 (free)   NASA events and trivia for each day of the year. See what happened in space on your birthday.

Appendix IV

Quotable Space Facts and Stats

SPACE TERRAIN

- If you look at some maps of space between the Sun and Mars, with Earth in between, the blackness of space is intersected with curved lines that look a lot like a topographic map of a wilderness on Earth. (The website **ScienceArt.com** contains some of the most beautiful images.)

- Lagrangian points are the "flat" places in space where the pulling and pushing forces are in balance. The orbiting object constantly pushes away from one body, held back by gravity, while the pulling of a second body keeps the object in the same place relative to the two large bodies. (Paul A. Tipler, *Elementary Modern Physics*, Chapter 9: Astrophysics and Cosmology)

- Each pair of bodies in the Solar System has five Lagrangian points between them. (Space.com has a good review of our Solar System's Lagrangian points. Search the site for "LaGrange Points." Also see NASA Earth Observatory, **EarthObservatory.nasa.gov**.)

- Venus used to be a pleasant, watery, habitable planet much like ours. The Martian atmosphere, it turns out, also used to be much more like Earth's. Both planets built up carbon dioxide, trapping the Sun's rays and causing the planets to warm dramatically. (Fredric W. Taylor, *The Scientific Exploration of Venus*; Giles Sparrow, *Mars*)

THE GRAVITY WELL

- Most of today's space economy occupies low Earth orbit, a 1,200-mile-wide band that's a slice of less than one one-thousandth of the Gravity Well. (Wikipedia has current information on satellites in LEO, MEO, and GEO. Search for "satellites.")

- To get into low Earth orbit—a path in which the forces of motion balance the pull of gravity—a rocket has to attain a speed of about 17,000 miles per hour. To escape the Gravity Well, a vehicle must go about 25,000 miles per hour—33 times the speed of sound. (Daniel Fleisch and Julia Kregenow, *A Student's Guide to the Mathematics of Astronomy*)

- As a rule of thumb, ten pounds of rocket fuel are required to push one pound of equipment or human into low Earth orbit. (Howard E. McCurdy, *Faster, Better, Cheaper: Low-Cost Innovation in the U.S. Space Program*)

- At 240,000 miles on average, the Moon lies only one fifth of the way to the top of the Gravity Well. This is equal to about 80 Columbus voyages. But you need far more energy to get from Earth to the Moon than from the Moon to Mars. (Peter Cadogan, *The Moon—Our Sister Planet*)

- To travel to Mars, astronauts will have to complete about 8,750 Columbus voyages of distance, assuming that Mars and Earth are at their closest points in their respective orbits around the Sun, a position they take once every two years. While the distance is vast, the fuel cost is not as much greater as it looks; the terrain from the Moon to Mars is milder than that from Earth to the Moon. (Calculation based on Columbus's first voyage, from his journal, edited by Clements R. Markham.)

HISTORY

- The first commercial flight took place on New Year's Day, 1914, with a ticket price of $400. It carried a single passenger from St. Petersburg to Tampa, Florida. (Tim Sharp, "World's First Commercial Airline," in **space.com**)

- In April, 2001, Space Adventures sent American businessman Dennis Tito to the International Space Station for a reported $20 million payment, making him the first space tourist.

- At the height of the attempt to send men to the Moon— when NASA sent three humans 25,000 miles an hour into space, circled the Moon, and brought them back safely— most citizens opposed the program. (Herbert E. Krugman, "Public Attitudes Toward the Apollo Space Program, 1965– 1975," *Journal of Communication,* December 1977)

- On July 20, 1969, an estimated 600 million people around the globe watched Neil Armstrong take his first steps onto the lunar surface. This was about 15% of everyone alive at that time. ("When the Eagle Landed," *Wall Street Journal,* July 16, 2009)

- Three years passed between the first transcontinental airmail service and the establishment of the first transcontinental airline. ("Airmail creates an industry." *National Postal Museum*, Smithsonian, 2004; "From mail-sack seats to sleeping berths and above-cloud routes," *Boeing Frontiers*.)

- Twelve years passed between the Soviets' first rocket and the Americans' Moon landing. (The R-7 intercontinental missile launched in 1957. The Moon landing took place in 1969.)

- Forty-four years have passed since the last time humans were on the Moon. (Eugene Cernan, commander of Apollo 17, holds the distinction of being the last person to set foot on the Moon, December 14, 1972. See the Smithsonian Air and Space Museum website, **AirandSpace.si.edu**.)

- The five Space Shuttles completed 135 missions over 22 years, until the program ended in 2011. (Piers Bizony, *The Space Shuttle*)

- Sixteen nations, including America, participated in building the International Space Station. (Conclusion in W. Henry Lambright, ed., *Space Policy in the 21st Century*.)

NASA

- America's space agency isn't just about space. Its major initiatives comprise astrophysics, the purest science of the cosmos; study of our Solar System; aeronautics, focusing on improvements to aviation technology; Earth science; examination of the Sun; space technology, which finds new ways to conduct robotic and human missions; and, finally, the human exploration program. (**nasa.gov**)

• To bring humans and their equipment into space, NASA is building its biggest rocket ever, the space launch system or SLS. The first version will push 77 tons into orbit. Eventually, the plan is to build a rocket that can carry 143 tons into space—10% more thrust than the Saturn V. (**AerospaceGuide.net**)

• When Eisenhower created NASA, the highest earners in America were paying a top marginal tax rate of more than 90%; it is now just over 39%. (**TaxFoundation.org**)

• Over the previous three decades, our nation's leaders have reduced NASA's buying power by 25%. (TheGuardian. com, Data Blog, "NASA Budgets: U. S. spending on space travel since 1958 updated"; contains a chart showing budgets as well as NASA's portion of the federal budget.)

• The 2015 space movie, *The Martian*—the one in which Matt Damon plays an astronaut-botanist—took in more box office receipts on its opening weekend than NASA's daily budget. (**BoxOfficeMojo.com**. The movie took in $54.3 million on its first weekend. NASA's daily budget is $52.8 million.)

• The amount NASA needs is comparable to someone with an income of $60,000 spending $10 annually on space. (This rough estimate assumes a $30 billion NASA budget. Individual income taxes provide 47% of federal revenue, according to the Center on Budget and Policy Priorities. A person with a $60,000 income, no children, and no special circumstances pays $8,196.25, according to **TaxFormCalculator.com**.)

ECONOMY

- Aviation contributes $1.5 trillion—more than 5%—to the national economy. (Federal Aviation Administration, "The Economic Impact of Civil Aviation on the U.S. Economy," June 2014)

- The Apollo program led to the creation of the light-weight mini-computer, GPS, the kidney dialysis machine, the smoke detector, the rechargeable pacemaker, the CPAP breathing machine, advanced prosthetics, a lead paint detector, improved vehicle brakes, freeze-dried food, and camera-on-a-chip technology used in smartphones today. (**Spinoff.nasa.gov**)

- More than a thousand satellites occupy low Earth orbit, or LEO. (Programmer James Yoder has created a website that lets you see everything orbiting the Earth, including satellites and junk, in real time: **StuffIn.space**.)

- Several dozen satellites are in medium Earth orbit, or MEO, providing our GPS. (For a complete list of individual satellites, see **Satellites.FindtheData.com**.)

• More than 250 satellites in geosynchronous orbit, or GEO, enable our Internet, television, and telephone communications. (Wikipedia has a current list of satellites in GEO. Search for "list of satellites in geosynchronous orbit.")

• The space industry generates more than $300 billion in revenue, an amount that continues to grow 5% to 15% a year. (Satellite Industry Association, **sia.org**)

• For every dollar spent on space, the knowledge ore brings back as much as $8. And that's not counting the 65,000 jobs created by the space program—which uses private contractors to do most of the work. (The multiplier is an extrapolation from Jerome Schnee, "The Economic Impacts of the U.S. Space Program," available at er.jsc.nasa.gov.) The jobs figure is a conservative estimate, starting with the 18,000 NASA employees (**nasa.gov**) and tens of thousands of on-site contactors at NASA's Centers.

• Satellite-based weather tracking has an annual economic value of $11 billion. (Rodney F. Weiher, Chief Economist, NOAA)

• One asteroid can hold $20 trillion worth of minerals, more than the annual gross domestic product of the American economy. A small, 100-foot asteroid can contain as much as $50 billion of platinum. (For a ranking of the value and mineral content of more than 600,000 asteroids, see **asterank.com**.)

PRIVATE SECTOR

• Elon Musk put up $100 million of his own money for SpaceX and received $1.25 billion in venture capital. NASA has awarded the company $3 billion in contracts. ("SpaceX overview on secondmarket," Secondmarket, a division of NASDAQ)

• It costs $50 million to $500 million to put a satellite into low Earth orbit. SpaceX promises a price per pound in low Earth orbit of $709—one-seventh the current price. (SpaceX.com. The company's stated cost of a Falcon Heavy rocket is $90 million; it carries a payload into low Earth orbit of almost 120,000 pounds. Motley Fool, fool.com, contains a list of commercial companies and the cost of launching satellites with each. Search for "How Much Does It Cost to Launch a Satellite?")

• Sixteen private companies compete in space, along with three commercial wings of national space agencies. (Wikipedia has an up-to-date list of companies. Search for "List of private space companies.")

• The Apollo mission directed 85% of its budget—more than $100 million—to private companies. That percentage holds true to this day. (**space.com**)

• The private sector funds just 18% of basic research in this country. The public sector funds 57%, with universities and foundations covering the rest. (Mariana Mazzucato, *The Entrepreneurial State.* Also see the MIT report, "The

Future Postponed: Why Declining Investment in Basic Research Threatens a U.S. Innovation Deficit," reported in **BusinessInsider.com**.)

• Federal employees comprise just 20% of the total NASA mission workforce. ("Employee Orientation," **nasa.gov**)

STEM

• From 2004 to 2014, the overall job market grew by 4.5%, while tech jobs expanded by 31%. (Bureau of Labor Statistics)

• The median annual salary for STEM jobs in 2014 was $78,610, more than double the $33,900 median salary for non-STEM; the mean salary for STEM was $85,570, and it was $47,230 for non-STEM. (Burning Glass Technologies, burning-glass.com. Also see *Forbes* magazine, "The Valley and the Upstarts.")

• Only 16% of American high school students are interested in a STEM career and proficient in math. Of the students who pursue a college major in a STEM discipline, only about half decide to work in a STEM career. The United States ranks a dismal 34th among industrialized nations in math, and 27th in science. (*U.S. News*, "Behind America's Decline in Math, Science, and Technology")

• The White House projects that by 2018 there will be 2.4 million unfilled STEM jobs in the United States. While an estimated 1.4 million U.S. science-related jobs will exist by 2020, American college graduates are expected to fill less than a third of them. (**WhiteHouse.gov**, "STEM Depiction Opportunities")

• Among top-performing high school students in 1992, just 29% went on to major in STEM subjects in college—far lower than in other developed nations. By the year 2000, the proportion of American STEM majors had dropped to 14%. By 2011, only 30% of high school seniors could even meet college enrollment standards in science. (*U.S. News*, "Behind America's Decline in Math, Science, and Technology")

• One out of 20 students in American colleges is a citizen of another country; yet more than half of all master's degrees and Ph.D.'s in STEM subjects go to foreign citizens. Non-Americans earn 57% of engineering doctoral degrees, 53% of computer and information sciences doctoral degrees, 50% of mathematics and statistics doctoral degrees, 49% of engineering-tech and engineering-related doctoral degrees, and 40% of doctorates in physical sciences and science technologies. (Pew Research Center/National Center for Education Statistics)

INTERNATIONAL

• More than 50 nations have their own satellites in low Earth orbit. (**n2yo.com** has a table of satellites by countries and organizations. Argentina, for example, has 14; Iraq, 1; and India, 70.)

- A third of the 1,469 satellites in orbit are American. (Union of Concerned Scientists, "USC Satellite Database," **ucsusa.org**)

Appendix V

James E. Webb
Letter to President Kennedy

November 30, 1962

This letter, written by NASA's legendary administrator, counts as one of the most prescient, still-relevant cases for maintaining all of NASA's missions. It follows a meeting with President Kennedy earlier in the month, when the President argued for speeding up the schedule for the Apollo Moon landing. The funds would come from other, non-Apollo, missions. Kennedy asked Webb to summarize his points; hence this letter, which ultimately convinced the boss. The Apollo schedule was maintained, and the science and the non-lunar technology missions remained intact. A decade later, Nixon approved the Space Shuttle program, but only at the expense of planetary science; and it took 30 years after the Viking landings in 1976 to get back to Mars.

The letter was published in Robert Seamans, Jr.'s excellent monograph, *Project Apollo: The Tough Decisions*.

NATIONAL AERONAUTICS AND SPACE ADMINISTRATION
WASHINGTON 25, D.C.
OFFICE OF THE ADMINISTRATOR

November 30, 1962

The President
The White House

Dear Mr. President:

At the close of our meeting on November 21,
concerning possible acceleration of the manned lunar
landing program, you requested that I describe for
you the priority of this program in our over-all
civilian space effort. This letter has been prepared
by Dr. Dryden, Dr. Seamans, and myself to express
our views on this vital question.

The objective of our national space program is to
become pre-eminent in all important aspects of this
endeavor and to conduct the program in such a manner
that our emerging scientific, technological, and
operational competence in space is clearly evident.

To be pre-eminent in space, we must conduct
scientific investigations on a broad front. We must
concurrently investigate geophysical phenomena about
Earth, analyze the sun's radiation and its effect
on Earth, explore the Moon and the planets, make
measurements in interplanetary space, and conduct
astronomical measurements.

To be pre-eminent in space, we must also have an advancing technology that permits increasingly large payloads to orbit Earth and to travel to the Moon and the planets. We must substantially improve our propulsion capabilities, must provide methods for delivering large amounts of internal power, must develop instruments and life support systems that operate for extended periods, and must learn to transmit large quantities of data over long distances.

To be pre-eminent in operations in space, we must be able to launch our vehicles at prescribed times. We must develop the capability to place payloads in exact orbits. We must maneuver in space and rendezvous with cooperative spacecraft and, for knowledge of the military potentials, with uncooperative spacecraft. We must develop techniques for landing on the Moon and the planets, and for re-entry into Earth's atmosphere at increasingly high velocities. Finally, we must learn the process of fabrication, inspection, assembly, and check-out that will provide vehicles with life expectancies in space measured in years rather than months. Improved reliability is required for astronaut safety, long duration scientific measurements, and for economical meteorological and communications systems.

In order to carry out this program, we must continually up-rate the competence of Government research and flight centers, industry, and universities, to implement their special assignments and to work together effectively toward common goals.

We also must have effective working relationships with many foreign countries in order to track and acquire data from our space vehicles and to carry out research projects of mutual interest and to utilize satellites for weather forecasting and world-wide communications.

Manned Lunar Landing Program

NASA has many flight missions, each directed toward an important aspect of our national objective. The manned lunar landing program requires for its successful completion many, though not all, of these flight missions. Consequently, the manned lunar landing program provides currently a natural focus for the development of national capability in space and, in addition, will provide a clear demonstration to the world of our accomplishments in space. The program is the largest single effort within NASA, constituting three-fourths of our budget, and is being executed with the utmost urgency. All major activities of NASA, both in headquarters and in the field, are involved in this effort, either partially or full time.

In order to reach the Moon, we are developing a launch vehicle with a payload capability 85 times that of the present Atlas booster. We are developing flexible manned spacecraft capable of sustaining a crew of three for periods up to 14 days. Technology is being advanced in the areas of guidance and navigation, re-entry, life support, and structures-in short, almost all elements of booster and spacecraft technology.

The lunar program is an extrapolation of our Mercury experience. The Gemini spacecraft will provide the answers to many important technological problems before the first Apollo flights. The Apollo program will commence with Earth orbital maneuvers and culminate with the one-week trip to and from the lunar surface. For the next five to six years there will be many significant events by which the world will judge the competence of the United States in space.

The many diverse elements of the program are now being scheduled in the proper sequence to achieve this objective and to emphasize the major milestones as we pass them. For the years ahead, each of these tasks must be carried out on a priority basis.

Although the manned lunar landing requires major scientific and technological effort, it does not encompass all space science and technology, nor does it provide funds to support direct applications in meteorological and communications systems. Also, university research and many of our international projects are not phased with the manned lunar program, although they are extremely important to our future competence and posture in the world community.

Space Science

As already indicated, space science includes the
following distinct areas: geophysics, solar physics,
lunar and planetary science, interplanetary science,
astronomy, and space biosciences.

At present, by comparison with the published
information from the Soviet Union, the United States
clearly leads in geophysics, solar physics, and
interplanetary science. Even here, however, it must
be recognized that the Russians have within the
past year launched a major series of geophysical
satellites, the results of which could materially
alter the balance. In astronomy, we are in a period
of preparation for significant advances, using the
Orbiting Astronomical Observatory which is now under
development. It is not known how far the Russian
plans have progressed in this important area. In
space biosciences and lunar and planetary science,
the Russians enjoy a definite lead at the present
time. It is therefore essential that we push forward
with our own programs in each of these important
scientific areas in order to retrieve or maintain
our lead, and to be able to identify those areas,
unknown at this time, where an added push can make a
significant breakthrough.

A broad-based space science program provides
necessary support to the achievement of manned space
flight leading to lunar landing. The successful
launch and recovery of manned orbiting spacecraft
in Project Mercury depended on knowledge of the

pressure, temperature, density, and composition
of the high atmosphere obtained from the nation's
previous scientific rocket and satellite program.
Considerably more space science data are required
for the Gemini and Apollo projects. At higher
altitudes than Mercury, the spacecraft will approach
the radiation belt through which man will travel
to reach the Moon. Intense radiation in this belt
is a major hazard to the crew. Information on
the radiation belt will determine the shielding
requirements and the parking orbit that must be used
on the way to the Moon.

Once outside the radiation belt, on a flight to
the Moon, a manned spacecraft will be exposed to
bursts of high speed protons released from time
to time from flares on the sun. These bursts do not
penetrate below the radiation belt because they are
deflected by Earth's magnetic field, but they are
highly dangerous to man in interplanetary space.

The approach and safe landing of manned
spacecraft on the Moon will depend on more precise
information on lunar gravity and topography.
In addition, knowledge of the bearing strength
and roughness of the landing site is of crucial
importance, lest the landing module topple or sink
into the lunar surface.

Many of the data required for support of the
manned lunar landing effort have already been
obtained, but as indicated above there are many
crucial pieces of information still unknown. It is
unfortunate that the scientific program of the past

decade was not sufficiently broad and vigorous to have provided us with most of these data. We can learn a lesson from this situation, however, and proceed now with a vigorous and broad scientific program not only to provide vital support to the manned lunar landing, but also to cover our future requirements for the continued development of manned flight in space, for the further exploration of space, and for future applications of space knowledge and technology to practical uses.

Advanced Research and Technology

The history of modern technology has clearly shown that pre-eminence in a given field of endeavor requires a balance between major projects which apply the technology, on the one hand, and research which sustains it on the other. The major projects owe their support and continuing progress to the intellectual activities of the sustaining research. These intellectual activities in turn derive fresh vigor and motivation from the projects. The philosophy of providing for an intellectual activity of research and an interlocking cycle of application must be a cornerstone of our National Space Program.

The research and technology information which was established by the NASA and its predecessor, the NACA, has formed the foundation for this nation's pre-eminence in aeronautics, as exemplified by our military weapons systems, our world market in civil jet airliners, and the unmatched manned flight

within the atmosphere represented by the X-15. More
recently, research effort of this type has brought
the TFX concept to fruition and similar work will
lead to a supersonic transport which will enter
a highly competitive world market. The concept
and design of these vehicles and their related
propulsion, controls, and structures were based on
basic and applied research accomplished years ahead.
Government research laboratories, universities, and
industrial research organizations were necessarily
brought to bear over a period of many years prior to
the appearance before the public of actual devices
or equipment.

These same research and technological manpower
and laboratory resources of the nation have formed
a basis for the U.S. thrust toward pre-eminence
in space during the last four years. The launch
vehicles, spacecraft, and associated systems
including rocket engines, reaction control systems,
onboard power generation, instrumentation and
equipment for communications, television and the
measurement of the space environment itself have
been possible in this time period only because
of past research and technological effort. Project
Mercury could not have moved as rapidly or as
successfully without the information provided by
years of NACA and later NASA research in providing
a base of technology for safe re-entry heat shields,
practical control mechanisms, and life support
systems.

It is clear that a pre-eminence in space in the future is dependent upon an advanced research and technology program which harnesses the nation's intellectual and inventive genius and directs it along selective paths. It is clear that we cannot afford to develop hardware for every approach but rather that we must select approaches that show the greatest promise of payoff toward the objectives of our nation's space goals. Our research on environmental effects is strongly focused on the meteoroid problem in order to provide information for the design of structures that will insure their integrity through space missions. Our research program on materials must concentrate on those materials that not only provide meteoroid protection but also may withstand the extremely high temperatures which exist during re-entry as well as the extremely low temperatures of cryogenic fuels within the vehicle structure. Our research program in propulsion must explore the concepts of nuclear propulsion for early 1970 applications and the even more advanced electrical propulsion systems that may become operational in the mid-1970's. A high degree of selectivity must be and is exercised in all areas of research and advanced technology to ensure that we are working on the major items that contribute to the nation's goals that make up an over-all pre-eminence in space exploration. Research and technology must precede and pace these established goals or a stagnation of progress in space will inevitably result.

Space Applications

The manned lunar landing program does not
include our satellite applications activities.
There are two such program areas under way and
supported separately: meteorological satellites
and communications satellites. The meteorological
satellite program has developed the TIROS system,
which has already successfully orbited six
spacecraft and which has provided the foundation
for the joint NASA-Weather Bureau planning
for the national operational meteorological
satellite system. This system will center on the
use of the Nimbus satellite which is presently
under development, with an initial research and
development flight expected at the end of 1963. The
meteorological satellite developments have formed an
important position for this nation in international
discussions of peaceful uses of space technology for
world benefits.

NASA has under way a research and development
effort directed toward the early realization of
a practical communication satellite system. In
this area, NASA is working with the Department of
Defense on the Syncom (stationary, 24-hour orbit,
communications satellite) project in which the
Department of Defense is providing ground station
support for NASA's spacecraft development; and
with commercial interests, for example, AT&T on
the Telstar project. The recent "Communications
Satellite Act of 1962" makes NASA responsible

for advice to and cooperation with the new
Communications Satellite Corporation, as well as
for launching operations for the research and/
or operational needs of the Corporation. The
details of such procedures will have to be defined
after the establishment of the Corporation. It is
clear, however, that this tremendously important
application of space technology will be dependent
on NASA's support for early development and
implementation.

University Participation

In our space program, the university is the
principal institution devoted to and designed for
the production, extension, and communication of new
scientific and technical knowledge. In doing its
job, the university intimately relates the training
of people to the knowledge acquisition process of
research. Further, they are the only institutions
which produce more trained people. Thus, not only
do they yield fundamental knowledge, but they are
the sources of the scientific and technical manpower
needed generally for NASA to meet its program
objectives.

In addition to the direct support of the space
program and the training of new technical and
scientific personnel, the university is uniquely
qualified to bring to bear the thinking of
multidisciplinary groups on the present-day problems
of economic, political, and social growth. In this

regard, NASA is encouraging the universities to
work with local industrial, labor, and governmental
leaders to develop ways and means through which the
tools developed in the space program can also be
utilized by the local leaders in working on their
own growth problems. This program is in its infancy,
but offers great promise in the working out of new
ways through which economic growth can be generated
by the spin-off from our space and related research
and technology.

International Activity

The National Space Program also serves as the base
for international projects of significant technical
and political value. The peaceful purposes of
these projects have been of importance in opening
the way for overseas tracking and data acquisition
sites necessary for manned flight and other programs
which, in many cases, would otherwise have been
unobtainable. Geographic areas of special scientific
significance have been opened to cooperative sounding
rocket ventures of immediate technical value. These
programs have opened channels for the introduction
of new instrumentation and experiments reflecting
the special competence and talent of foreign
scientists. The cooperation of other countries-
indispensable to the ultimate achievement of
communication satellite systems and the allocation
of needed radio frequencies-has been obtained in
the form of overseas ground terminals contributed

by those countries. International exploitation
and enhancement of the meteorological experiments
through the synchronized participation of some
35 foreign nations represent another by-product
of the applications program and one of particular
interest to the less developed nations, including
the neutrals, and even certain of the Soviet bloc
satellite nations.

These international activities do not in
most cases require special funding; indeed, they
have brought participation resulting in modest
savings. Nevertheless, this program of technical
and political value can be maintained only as an
extension of the underlying on-going programs, many
of which are not considered part of the manned lunar
landing program, but of importance to space science
and direct applications.

Summary and Conclusion

In summarizing the views which are held by Dr.
Dryden, Dr. Seamans, and myself, and which have
guided our joint efforts to develop the National
Space Program, I would emphasize that the manned
lunar landing program, although of highest national
priority, will not by itself create the pre-
eminent position we seek. The present interest
of the United States in terms of our scientific
posture and increasing prestige, and our future
interest in terms of having an adequate scientific
and technological base for space activities beyond

the manned lunar landing, demand that we pursue an adequate, well-balanced space program in all areas, including those not directly related to the manned lunar landing. We strongly believe that the United States will gain tangible benefits from such a total accumulation of basic scientific and technological data as well as from the greatly increased strength of our educational institutions. For these reasons, we believe it would not be in the nation's long-range interest to cancel or drastically curtail on-going space science and technology development programs in order to increase the funding of the manned lunar landing program in fiscal year 1963.

The fiscal year 1963 budget for major hardware development and flight missions not part of the manned lunar landing program, as well as the university program, totals $400 million. This is the amount which the manned space flight program is short. Cancellation of this effort would eliminate all nuclear developments, our international sounding rocket projects, the joint U.S.-Italian San Marcos project recently signed by Vice President Johnson, all of our planetary and astronomical flights, and the communication and meteorological satellites. It should be realized that savings to the Government from this cancellation would be a small fraction of this total since considerable effort has already been expended in fiscal year 1963. However, even if the full amount could be realized, we would strongly recommend against this action.

In aeronautical and space research, we now have a program under way that will insure that we are covering the essential areas of the "unknown." Perhaps of one thing only can we be certain; that the ability to go into space and return at will increases the likelihood of new basic knowledge on the order of the theory that led to nuclear fission.

Finally, we believe that a supplemental appropriation for fiscal year 1963 is not nearly so important as to obtain for fiscal year 1964 the funds needed for the continued vigorous prosecution of the manned lunar landing program ($4.6 billion) and for the continuing development of our program in space science ($670 million), advanced research and technology ($263 million), space application ($185 million), and advanced manned flight including nuclear propulsion ($485 million). The funds already appropriated permit us to maintain a driving, vigorous program in the manned space flight area aimed at a target date of late 1967 for the lunar landing. We are concerned that the efforts required to pass a supplemental bill through the Congress, coupled with Congressional reaction to the practice of deficiency spending, could adversely affect our appropriations for fiscal year 1964 and subsequent years, and permit critics to focus on such items as charges that "overruns stem from poor management" instead of on the tremendous progress we have made and are making.

As you know, we have supplied the Bureau of
the Budget complete information on the work that
can be accomplished at various budgetary levels
running from $5.2 billion to $6.6 billion for
fiscal year 1964. We have also supplied the Bureau
of the Budget with carefully worked out schedules
showing that approval by you and the Congress of
a 1964 level of funding of $6.2 billion together
with careful husbanding and management of the
$3.7 billion appropriated for 1963 would permit
maintenance of the target dates necessary for the
various milestones required for a final target date
for the lunar landing of late 1967. The jump from
$3.7 billion for 1963 to $6.2 billion for 1964 is
undoubtedly going to raise more questions than
the previous year jump from $1.8 billion to $3.7
billion.

If your budget for 1964 supports our request
for $6.2 billion for NASA, we feel reasonably
confident we can work with the committees and leaders
of Congress in such a way as to secure their
endorsement of your recommendation and the incident
appropriations. To have moved in two years from
President Eisenhower's appropriation request for
1962 of $1.1 billion to the approval of your own
request for $1.8 billion, then for $3.7 billion for
1963 and on to $6.2 billion for 1964 would represent
a great accomplishment for your administration.
We see a risk that this will be lost sight of in
charges that the costs are sky-rocketing, the

program is not under control, and so forth, if we request a supplemental in fiscal year 1963.

However, if it is your feeling that additional funds should be provided through a supplemental appropriation request for 1963 rather than to make the main fight for the level of support of the program on the basis of the $6.2 billion request for 1964, we will give our best effort to an effective presentation and effective use of any funds provided to speed up the manned lunar program.

With much respect, believe me
Sincerely yours,

James E. Webb
Administrator

Appendix VI

NASA Organization

OFFICE OF THE ADMINISTRATOR

ADVISORY GROUPS

NASA Advisory Council (NAC)

Aerospace Safety Advisory Panel (ASAP)

OFFICE OF THE INSPECTOR GENERAL

MISSION DIRECTORATES

Aeronautics Research

Human Exploration and Operations

Science

Space Technology

CENTERS AND FACILITIES

Ames Research Center   IT, fundamental aeronautics, bio and space science technologies

Armstrong Flight Research Center   Flight research

Glenn Research Center   Aero propulsion and communications technologies

257

Goddard Space Flight Center Earth, the Solar System, and Universe observations

Headquarters Agency leadership

Jet Propulsion Laboratory Robotic exploration of the Solar System

Johnson Space Center Human space exploration

Kennedy Space Center Prepare and launch missions around Earth and beyond

Langley Research Center Aeronautics and space research

Marshall Space Flight Center Space transportation and propulsion technologies

Stennis Space Center Rocket propulsion testing and remote sensing technology

Goddard Institute for Space Studies Broad study of global climate change

Independent Verification and Validation Facility Provides safety and cost-effectiveness for mission critical software

Michoud Assembly Facility Manufacture and assembly of critical hardware for exploration vehicles

NASA Engineering and Safety Center Independent testing, analysis, and assessments of NASA's high-risk projects

NASA Safety Center Development of personnel, processes and tools needed for the safe and successful achievement of strategic goals

NASA Shared Service Center   Financial management, human resources, information technology, and procurement

Wallops Flight Facility   Suborbital Research Programs

Appendix VII

Space in the States

$\overline{H}$ere is a sampling of space's presence throughout the nation. NASA's Space Grant Program has established 52 consortia— networks of colleges and universities—in every state plus the District of Columbia and Puerto Rico. These consortia support education, research, and outreach for space. Some individual state consortia are listed below. See **TheGravityWell.org** for an updated list and links.

ALABAMA

NASA Marshall Space Flight Center   NASA's principal workshop for developing transportation and propulsion technologies.

U.S. Army Redstone Test Center   A unit of the U.S. Army responsible for testing aviation technology, rockets, and sensors.

ALASKA

Alaska Satellite Facility   Part of the University of Alaska Fairbanks' Geophysical Institute, this satellite-tracking ground station processes remote sensing data to facilitate climate research.

Eareckson Air Station   Landing and refueling stop for NASA aircraft.

Kodiak Launch Complex   Commercial rocket launch facility for suborbital and orbital space launch vehicles owned by the Alaska Aerospace Corporation.

ARIZONA

Air Force Research Laboratory, Mesa Air Force Base
Arizona-based unit of the AFRL, which has helped NASA develop space vehicles, satellites, and lasers.

NASA Planetary Aeolian Laboratory, ASU Tempe
A research facility specializing in the study of windblown particles in various planetary atmospheres.

NASA Ronald Greeley Center for Planetary Studies, ASU Tempe Facility tasked with producing planetary images for use in educational communities.

ARKANSAS

Arkansas Center for Space and Planetary Sciences, University of Arkansas Institute housing departments that specialize in planetary sciences, astrophysics, and aerospace engineering, among other areas.

Arkansas Space Grant Consortium A network of 17 four-year colleges and universities familiarizing students with aerospace industries and promoting the involvement of the state of Arkansas in NASA activities.

CALIFORNIA

Air Force Research Laboratory, Edwards Air Force Base
California-based unit that has helped NASA test engines and rocket propulsion.

NASA Ames Research Center Focuses on human space travel and quality of life in the aerospace exploration age.

NASA Armstrong Flight Research Center   Formerly known as the Dryden Flight Research Center, this aeronautical research hub develops NASA's most advanced vehicles.

NASA Jet Propulsion Laboratory   Managed by the nearby Caltech, this NASA R&D center focuses on the construction and operation of planetary robotic spacecraft.

SpaceX   Elon Musk's rocket and spacecraft manufacturing and transportation company.

Vandenberg Air Force Base Space Launch Complex 6 Rocket launch pad and support site originally developed for NASA's Titan III and Manned Orbiting Laboratory missions.

Virgin Galactic   Sir Richard Branson's private spaceflight company, which is currently developing the technology to facilitate space tourism.

COLORADO

Air Force Space Command, Peterson Air Force Base A command site that supports military satellite launches and operations, with the help of more than 80 remote sites worldwide.

Ball Aerospace & Technologies   R&D giant producing space vehicles along with civil space and commercial space components.

Lockheed Martin Space Systems   Headquarters of one of Lockheed's four major business divisions, famous for developing NASA's Orion Multi-Purpose Crew Vehicle.

United Launch Alliance   Another aerospace launch vehicle behemoth, a joint venture between Lockheed Martin Space Systems and Boeing Defense, Space & Security.

Sierra Nevada Corporation Space Systems Makes four product lines, including spacecraft actuators for Mars rovers, hybrid rocket technologies, microsatellites controlled by the Internet, along with a winged and piloted orbital commercial spacecraft.

UP Aerospace Private spaceflight corporation that regularly delivers payloads for corporate, education, and military clients.

CONNECTICUT

Connecticut Space Grant Consortium One of NASA's 52 space grant consortium facilities, established to provide support to local space research initiatives.

DELAWARE

The Delaware AeroSpace Education Foundation Nonprofit organization devoted to promoting and facilitating STEM education subjects, including space and engineering.

DISTRICT OF COLUMBIA

NASA Headquarters Where NASA leadership and administrative executives convene. Also the primary hosting site for NASA press conferences.

FLORIDA

Florida Space Institute   Established by the state of Florida in 1996, this center, a regular NASA partner, specializes in planetary science and space weather.

NASA Cape Canaveral Air Force Station   Launch site for some of NASA's most important missions, including John Glenn's historic flight around the Earth.

NASA Kennedy Space Center   Origin point for every one of NASA's human space flights since 1968.

Space Florida   Official aerospace economic development agency of the state of Florida.

GEORGIA

Georgia Center of Innovation for Aerospace   Aerospace think tank that provides intel support to statewide aerospace companies.

Georgia Tech Center for Space Technology and Research   Hub of Georgia Tech's space science and engineering research activities. A regular partner with NASA.

HAWAII

Hawaii Space Flight Laboratory, University of Hawaii at Manoa   Multidisciplinary education and research center, specializing in spacecraft development and space workforce training.

Mauna Kea Observatories, Big Island of Hawaii   A 525-acre special land use zone on and around the summit of this volcano, where 13 telescopes are located.

IDAHO

Center for Space Nuclear Research   Joint venture between Idaho National Laboratories and the Universities Space Research Association, focusing on nuclear-powered space systems.

Northwest Nazarene University   Launch site for NASA's CubeSat Launch Initiative missions, sending small satellites into space.

ILLINOIS

Illinois Space Society, University of Illinois at Champaign-Urbana   Educational outreach group that connects students and professionals with space research opportunities.

Illinois Space Grant Consortium   Consortium of educational institutions across the State of Illinois focused on the education of the next generation of aerospace engineers and scientists through scholarships, fellowships, hands-on projects, course development, etc.

Boeing   100 years old in July 2016 and headquartered in Chicago, Boeing is the world's largest aerospace company and leading manufacturer of commercial jetliners and defense, space and security systems. A top U.S. exporter, the company supports airlines and U.S. and allied government customers in 150 countries.

INDIANA

Indiana Space Grant Consortium   Promotes workforce development, STEM education and research. Includes colleges and universities, museums and science centers, and corporate affiliates.

Spaceport Indiana   Private company that offers space workshops to educational communities and assists industry researchers with data collection and evaluation.

IOWA

Doerfer Companies   Automotive developer behind the Wheelift Self-Propelled Modular Transporter, which NASA has used to move its massive SLS rockets.

KANSAS

Cosmosphere   Space museum and STEM education center featuring the largest collection of spacecraft artifacts in the world.

Kalscott Engineering   R&D firm that designs flight systems; a recent partner with NASA's Small Business Innovation Research/Small Business Technology Transfer program.

KENTUCKY

Space Science Center, Morehead State University Division of the university's science department that specializes in aerospace systems development.

269

Space Systems Laboratory, University of Kentucky
Student-majority developer of small payload satellite
technologies and a collaborator with NASA.

LOUISIANA

NASA Michoud Assembly Facility Where NASA
constructs its rockets and spacecraft, in some of the largest
assembly structures in the world.

MAINE

Fiber Materials, Inc. Developer of high temperature
resistant composites, insulation, and other materials routinely
used in aerospace missions.

Maine Aerospace Consulting, LLC. Provides aerospace
and defense clients with algorithmic solutions to engineering
challenges.

MARYLAND

NASA Goddard Space Flight Center First NASA space
flight center, where the agency designs, builds, and tests
spacecraft technologies.

**The Johns Hopkins University Applied Physics
Laboratory** Located between Baltimore and Washington,
near I-95, and with many space flights including missions to
Mercury and to Pluto.

Space Telescope Science Institute Located at John
Hopkins University in Baltimore, this is the operations center
for the famous Hubble Space Telescope.

Maryland Space Grant Consortium Comprising 12 institutions, including the U.S. Naval Academy, distributed across the state.

MASSACHUSETTS

MIT Lincoln Laboratory Department of Defense R&D center specializing in space control systems; a NASA partner for environmental monitoring activities.

Raytheon Defense technology giant responsible for providing the engineering components behind NASA's training programs.

MICHIGAN

Michigan Space Grant Consortium Fosters awareness of, education in, and research on space-related science and technology in Michigan.

Precision Aerospace Corporation Provides small batches of high quality machined parts for commercial and defense aerospace companies.

Triple Inc. Manufactures precision components for the aerospace industry.

MINNESOTA

EXB Solutions Software company that provided NASA with the software and testing services necessary to launch the Orion spacecraft.

MISSISSIPPI

NASA Stennis Space Center   NASA's largest rocket engine test facility.

MISSOURI

Boeing Defense, Space & Security   Aerospace and defense headquarters of Boeing, located just outside Lambert-St. Louis International Airport.

MONTANA

Ascent Vision Technologies, LLC.   Designs precision technologies for manned and unmanned aeronautical crafts.

NEBRASKA

Lincoln Airport/Lincoln Air National Guard Base Alternate landing site for NASA missions.
University of Nebraska's College of Law   Home of the world's only English-instructed Master of Laws degree program in space, cyber and telecommunications.

NEVADA

Bigelow Aerospace   Space technology company that develops expandable modules for use in space.

Sierra Nevada Corporation   A privately held advanced technology company headquartered in Sparks, Nevada. Provides solutions in aerospace, aviation, electronics, and systems integration.

NEW HAMPSHIRE

Institute for the Study of Earth, Oceans and Space, University of New Hampshire   Develops aerospace instruments enabling NASA missions.

NEW JERSEY

Aerospace Manufacturing, Inc.   Aerospace and defense technology company known for its precision manufacturing.

Atlantic City International Airport   Emergency landing site for aborted NASA missions.

NEW MEXICO

Air Force Research Laboratory, Kirtland Air Force Base New Mexico based unit of the AFRL, which has helped NASA develop space vehicles.

Spaceport America   FAA-licensed spaceport from which tenants, including SpaceX, UP Aerospace, and Virgin Galactic, have conducted operations.

White Sands Test Facility   U.S. government testing ground for rocket engines, space flight components, and potentially hazardous materials.

NEW YORK

NASA Goddard Institute for Space Studies   Located at Columbia University, this research lab specializes in environmental changes that can affect the habitability of a planet.

New York Space Grant Consortium   Provides support for NASA-related STEM research and educational programs throughout New York State.

Ursa   Mapping and software company that converts satellite data into usable products.

Other companies that have space facilities in New York: BAE Systems, General Electric, Honeybee Robotics, Lockheed Martin, Moog, and Northrop Grumman.

NORTH CAROLINA

UTC Aerospace Systems   Internationally prolific developer of satellite technology and spacecraft components.

NORTH DAKOTA

10th Space Warning Squadron, Cavalier Air Force Station   One of the U.S. Air Force's many space surveillance sites, which use radar technology to identify space objects.

North Dakota Space Grant Consortium   Headquartered in the Space Studies Department at the University of North Dakota in Grand Forks. Includes agriculture, asteroid photometry and spectroscopy, and factors involved in human space flight.

OHIO

NASA Glenn Research Center   This NASA R&D site features a large-scale spacecraft propulsion testing facility, a zero gravity chamber, and a tunnel that mimics atmospheric icing.

OKLAHOMA

Clinton-Sherman Industrial Airpark   The primary asset of the Oklahoma Space Industry Development Authority, this relatively new spaceport has hosted aerospace test flights.

Oklahoma Air & Space Port   Public-use airport and industrial airpark that includes facilities for aerospace testing and launching. Boasts one of the country's longest and widest runways

OREGON

Near Space Corporation   This space flight integrations system manufacturer recently won a NASA contract to support the agency's Flight Opportunities Program.

PENNSYLVANIA

National Aerospace Training & Research Center   Non-governmental facility that provides training for pilots and supports aerospace research.

RHODE ISLAND

Northeast Planetary Data Center   Funded through an agreement between NASA and Brown University, the NEPDC is part of the international network of Regional Planetary Image Facilities housing vast reference collections of planetary images, data, and supporting documents for use by educators, students, and the public.

Rhode Island Space Grant Consortium   Congressionally mandated program to support NASA's missions and objectives through fellowships and scholarships, research seed awards, education partnerships, and higher education activities.

Rhode Island Aviation and Space Education Council This group was established to promote aviation education activities and programs to K-12 students.

SOUTH CAROLINA

Victory Solutions, Inc.   Houses the design-to-build site for this aerospace and defense engineering firm, which has partnered with NASA's Marshall Space Flight Center.

SOUTH DAKOTA

South Dakota School of Mines and Technology   This 100% STEM degree-granting university is working with NASA most recently to convert solid waste into a power source for long-term space missions.

TENNESSEE

Arnold Engineering Development Complex   Home to the largest complex of flight simulation test facilities in the world, including more than 50 wind tunnels.

TEXAS

Ellington Airport   Public and military airport that supports NASA operations.

Midland International Air and Space Port   Owned by the City of Midland, this facility was the first to receive an FAA license to serve commercial spaceflight and scheduled airline flights.

NASA Johnson Space Center   NASA field site specializing in human spaceflight support. Established in 1961 to facilitate Apollo.

SpaceX, Blue Origins and Armadillo Aerospace and XCOR have test facilities here.

UTAH

Space Dynamics Lab, Utah State University   Nonprofit research corporation that develops satellite and sensor technology and employs more than 400 scientists and engineers.

Orbital ATK has seven facilities in Utah including making and testing solid rocket booster engines.

Utah NASA Space Grant Consortium   Includes higher educational institutions, industry, government partners, a science center, and museums. Contributes to NASA's STEM enterprise by promoting education, research, precollege and informal education projects.

VERMONT

Revision Military Technologies   R&D firm (and NASA contractor) that develops laser-proof lenses designed to be worn by astronauts during automatic docking procedures.

VIRGINIA

Mid-Atlantic Regional Spaceport   One of only four spaceports licensed by the FAA to launch rockets into orbit. Conducts cargo resupply missions to the International Space Station.

NASA Langley Research Center   NASA's oldest field site. The majority of its research initiatives focus on atmospheric science and aeronautical challenges.

NASA Wallops Flight Facility   More than 16,000 rockets have been fired off from this NASA launch site, one of the oldest in the world. The facility has its own spaceport.

National Institute of Aerospace   Conducts aerospace and atmospheric R&D.

Northrop Grumman Corporation   One of the leading aerospace and defense technology companies in the world, and longtime contractor with NASA.

Orbital ATK   The aerospace manufacturer behind the Cygnus spacecraft, which has been used to transport supplies to the International Space Station.

Virginia Space Grant Consortium   Coalition of universities and state agencies together with NASA Langley

Research Center, NASA Wallops Flight Facility, and other organizations. Provides scholarships, fellowships, internships, student flight and design projects, student research and enrichment programs and faculty professional development and research.

WASHINGTON

Blue Origin   Aerospace manufacturer and spaceflight services company founded by Amazon's Jeff Bezos.

Boeing Everett Factory   The aerospace giant's Washington airplane assembly facility holds the distinction of being the biggest building in the world by volume.

Orion Industries   Prolific manufacturer of precision metal products used by the aerospace and defense industries.

Aerojet Rocketdyne   Provides propulsion and energetics to industry and government for aerospace.

Washington NASA Space Grant Consortium   One of 52 members of NASA's National Space Grant College and Fellowship Program, supporting space-related education and research in the Pacific Northwest.

Washington Aerospace Training & Research Center   Managed by Edmonds Community College, the center offers educational aerospace and manufacturing programs to students.

WEST VIRGINIA

NASA Independent Verification and Validation Facility Conducts rigorous safety tests for all mission software.

WISCONSIN

Orbital Technologies Corporation   Since 1988 this aerospace systems developer has won more than $210 million in combined contracts from NASA and other government and industry entities.

WYOMING

Frontier Astronautics, LLC.   Commercial rocket engine and attitude control systems developer, founded by a veteran of Lockheed Martin and NASA's Titan I and II programs.

Appendix VIII

Glossary

Aeronautics   The science of flight, including the physics and operation of flying vehicles.

Algae   A variety of photosynthetic organisms, some of which could provide oxygen and food on long-traveling spaceships.

Ames Research Center (ARC)   Conducts critical research and develops enabling technologies in astrobiology, information technology, fundamental space biology, nanotechnology, air traffic management, thermal protection systems and human factors essential to virtually all NASA missions. Based in Silicon Valley.

Arianespace   A French multinational corporation and commercial space service provider.

Armstrong Flight Research Center (ARFC)   Located within Edwards Air Force Base in southern California, Armstrong is NASA's lead for aeronautics.

Asteroid belt   Lying between Mars and Jupiter, this space contains a large percentage of the asteroids in our Solar System.

Astronomical unit (AU)   A standard space measurement based on the distance between Earth and the Sun—some 93 million miles.

Cosmology   The study of the universe, including its origins.

DOD   Department of Defense.

Deep Space Network (DSN)   A global communications network, including large antennas, to support missions beyond the Moon.

Energetic particle   Radiation capable of penetrating the walls of spacecraft, including gamma rays and X rays.

Escape velocity   The minimum velocity that allows an object or vehicle to escape gravity. The escape velocity for Earth's Gravity Well is 25,000 miles per hour.

European Space Agency (ESA)   A consortium consisting of Austria, Belgium, Denmark, Finland, France, Germany, Ireland, Italy, the Netherlands, Norway, Portugal, Spain, Sweden, Switzerland, and the United Kingdom. Canada is an affiliate.

Extravehicular Mobility Unit (EMU)   A two-piece spacesuit that allows an astronaut to live and communicate outside a spacecraft.

Gamma-ray burst (GRB)   A shower of high-energy radiation that generally lasts just a few seconds but could damage the DNA of astronauts outside Earth's magnetic field.

GEO   Geosynchronous or geostationary orbit, in which a satellite, vehicle or other object moves at the same speed as Earth's rotation. This "holds" the satellite in the same relative point above a spot on Earth. Weather satellites in GEO above Earth's equator "hover" at a distance of 22,300 miles above the surface.

Glenn Research Center (GRC)   An Ohio-based NASA center that transfers critical technologies that address national priorities through research, technology development, and systems development for safe and reliable aeronautics, aerospace, and space applications.

Goddard Space Flight Center (GSFC)   One of the chief NASA laboratories, as well as the control center for the Hubble Space Telescope and other systems. Goddard is located in Maryland outside of Washington, D.C.

Habitable zone   A region in space where planets contain liquid water, between the boiling and freezing points.

Hubble Space Telescope (HST)   A powerful telescope in low Earth orbit since 1990. Hubble boasts an eight-foot mirror as well as four instruments that measure light from celestial objects in visible, near-ultraviolet, and near-infrared wavelengths.

Intelsat   Originally an international organization of 11 nations, now a private company providing satellite service to more than 149 countries. Intelsat operates more than 50 communications satellites.

Inverse square law   Any relation where the force between objects (like gravity) decreases with the square of the distance between them. A spacecraft 6,000 miles from Earth experiences one quarter of the gravitational pull that it would 3,000 miles from the planet.

Jet Propulsion Laboratory (JPL)   One of NASA's chief centers, based in Pasadena, California, and operated by the California Institute of Technology (Caltech). JPL develops and operates robotic spacecraft, including the Mars rovers, among other projects.

Johnson Space Center (JSC)   The headquarters for human space flight, in Houston, Texas.

Kennedy Space Center (KSC)   Located on the East Coast of Florida, Kennedy is NASA's main launch center for human space flight. It also conducts launches for a variety of NASA missions.

Lagrangian point   A "flat" part of space terrain, where a satellite or station can remain in place relative to two bodies, without expending energy. Five such points exist between the Sun and Earth. L1 and L2, the most potentially useful points for space exploration, each lie a distance from Earth equal to four times the distance of the Moon.

Langley Research Center (LaRC)   Based in coastal Virginia, Langley served as the original laboratory for NACA, NASA's predecessor. Langley conducts aeronautics and space research for aerospace, atmospheric sciences, and technology commercialization.

LEO   Low Earth orbit, a region in space 89 to 1,200 miles above Earth.

Magnetosphere   A region around a planet where the magnetic field controls the motions of charged particles. Earth's magnetosphere deflects or traps radiation that would otherwise prove deadly to most species, including humans. Leaving the magnetosphere presents one of the great challenges to human space flight beyond the Gravity Well.

Marshall Space Flight Center (MSFC)   A NASA center in Huntsville, Alabama, that conducts research, designs and builds rockets, and manages space labs.

Medium Earth Orbit (MEO)   The region between low Earth orbit and geosynchronous orbit, 1,200 to 22,000 miles from Earth.

Microgravity   The condition of free fall, when an object appears to be weightless. Microgravity in space allows the formation of precise crystals, along with other forms of manufacture and experimentation that would be near-impossible on Earth.

NACA   The National Advisory Committee for Aeronautics, NASA's predecessor.

Optics   The branch of physics that studies the properties of light.

Speed of light   The speed at which photons move through empty space, about 186,000 miles per second.

Stennis Space Center   Stennis is responsible for NASA's rocket propulsion testing and for partnering with industry to develop and implement remote sensing technology.

Van Allen belts   A pair of high-radiation bands trapped by Earth's magnetic field, as close as 500 miles and as far as 40,000 miles from Earth.

Acknowledgments

While a book had not been one of my life's goals, I'm hoping that writing it will help me meet the bigger ones. Chief among them is my lifelong desire to solve the problem of expanding humanity beyond this planet. So these acknowledgments must include those who helped me explore this problem, long before I began working on the book.

My parents, Nancy and Maynard Sandford—two of the most remarkable people I've ever met—instilled in me the importance of getting to work on what needed doing. They taught me that life's riches lie in the people around us, and nothing else. My teachers in the public school system of Hampton, Virginia, were the best I possibly could have had. William Hesse, professor of physics at Randolph-Macon College, inspired me to pursue that discipline. Glenn Taylor hired me at NASA and mentored me at Langley Research Center and taught me to take on the hardest problems, working with the best experts wherever they might be found. A major thank you to the people I've worked with over the years at NASA, and especially at Langley Research Center. They can build anything that America chooses to build.

Jay Heinrichs helped crystallize the ideas in this book. He articulated the complex arguments more engagingly than I could have done alone, all while making the hard work seem fun. Jay Bennett, Lee Michaelides, and Miles Howard provided invaluable research on space history, and they fact checked the manuscript. Amy Herzog and Brad Herzog of Gavia Books copy edited the book and steered its way through the publishing process. Bobby Braun, Julie Witcover, Paul Witcover, Matt Yetman, Michelle Wilgenburg, and Mike Gazarik provided valuable comments on the manuscript.

Kam Ghaffarian deserves special thanks for his vision and support for this book.

Finally, I thank my friends, family, and colleagues who have listened to the ideas in this book and helped to improve them through their thinking and discussions over the years. My wife, Amy, and my children, Corin and Macey, are the reasons I get up every day. I thank them for their good ideas and for all those years of putting up with my goals and professional commitments.

Index

3D imaging detection devices, 120

technology transfer (T2) program,
122–124
terrain map of space, 29–33
terrorism, 152–155
Tito, Dennis, 225
tools-knowledge cycle. *See* invention-
discovery cycle
tourism, 77–79, 225
transportation system issues, 178
Turner, Frederick Jackson, 147

U
Udall, Mark, 35
uncertainty versus risk, 99–101
university enrollment, 137
university system, 15, 94, 98–99, 140
U.S. Post Office, 52–53, 226
USS *Ticonderoga*, 61

V
V-2 rocket, 53, 55, 57, 148–149
Valley of Death, 101–102
Van Allen radiation belts, 43
vault of heaven, 195–197
venture capitalists, 101–104
Venus, 37, 64, 223
vertical takeoff and landing (VTOL),
39, 80
Viking landers, 63–64
Virgin Galactic, 69, 77, 84
von Braun, Wernher, 55–56, 58, 62,
137, 148–149

W
Walcott, Charles D., 50–51
Walk & Roll Robot, 123
war as catalyst, 50–51
Watney, Mark, vii, 93
weather predictions, 41–43, 98, 229
Webb, James, 58, 235, 237–253
whispering galleries, 122
White Knight, 73

Wiesner, Jerome, 57
Wilson, Woodrow, 51
wind tunnels, 96–97
women in STEM, 134
wonder rooms, 145–146
World War I, 50–52, 54, 95, 98
World War II, 38, 52–54, 95, 138, 148
Wright brothers, 50, 62, 94, 148, 190
wunderkammern, 145–146

X
XCOR, 78
X-ray crystallography, 82

Y
Yeager, Chuck, 97
Yeltsin, Boris, 151
Yoder, James, 228

Z
zeppelins, 50
Zubrin, Robert, 116–117